Ginny Ho

Animal Commu:

Soul
Sketches

**Sketches from the soul of real
communications from real animals
and their humans**

Thank-you for Listening.

Ginny

x

EU Conformity Declaration

This product complies with the following safety regulations and standards to ensure consumer safety and product quality: Regulation (EU) 2023/988 of the European Parliament and of the Council on General Product Safety (GPSR): The Consumer Product Safety Improvement Act (CPSIA), Section 101. The Californian Safe drinking water and toxic enforcement act. (Proposition 65) EN71-Part 1: Mechanical and Physical Properties EN71-Part 2: Flammability EN71-Part 3 Migration of certain elements.

Published and Manufactured by Softwood Books
EU Responsible person: Maddy Glenn
Office 2, Wharfside House, Prentice Road, Stowmarket, Suffolk, IP14 1RD
www.softwoodbooks.com
hello@softwoodbooks.com

EU Rep:
Authorised Rep Compliance Ltd., Ground Floor, 71 Lower Baggot Street, Dublin, D02 P593, Ireland
www.arccompliance.com
info@arccompliance.com

Paperback ISBN: 978-1-0683776-0-0

Foreword

by Richard Geldard my teacher, my mentor, my friend.

I've known Ginny a few years now.

The first time I met her was on my Animal Communication course she attended.

From the moment I met her I could feel her warmth, her love, and her immense empathy towards animals, and it was very clear, very quickly, she had the gift also to be their voice.

Ginny has gone from strength to strength with the development, growth, and enlightenment of her gift, and it has been an absolute pleasure in guiding her in the direction that is clear — to anyone that meets her — she is so natural and fluent within.

One of the areas Ginny has always stood out to me in, is with her absolute desire and determination to make a difference to the horses of this world; to change people's perceptions of how they are treated, looked after, and understood.

I have also had the great pleasure of collaborating with Ginny on various projects, including her illustrating the video for my song *Whispers in the Wind*.

When Ginny talked to me about putting together a book like this, to me it completely made sense and resonated with me hugely, as along with her gift of communication, Ginny is without doubt a creative 'tour de force' who is instinctively fluent in illustrating the sheer feelings of the horse.

I am so excited that she has finally brought this book to life. Her journey continues.

Richard Geldard

International Animal Communicator, Medium & Spiritual teacher

Animals feel emotion as much as we do, and these sketches depict true stories which may touch you in places you didn't know you had. The scenarios have all been my observations and given to me by the animals to help teach us, and to help us understand them on a much deeper, more soulful level.

Real stories of real people I have helped, by being a voice for their animals, all from unconditional love and guided by Spirit.

Whilst the scenarios are real and have all been things I've seen, been told by the animals, have felt or interpreted, the characters are completely fictional. All my humans are faceless deliberately, so they could be anyone.

None of the depictions are coming from a critical place; they are all simply showing the truth in that moment, and unfortunately, the truth isn't always pretty.

They all have lessons within them and depending on how deep we dive in, will depend on the reflection and understanding we will gain from that experience which will be unique to every one of us.

We are all on our own individual journey and are coming to this with different learning experiences, conditioning, belief systems and personal experiences – but the messages from the animals will teach us all.

Self-reflection is key to these sketches, as is honesty and truth. The only way we can grow is to listen and acknowledge the lessons they provide, all from unconditional love.

As humans we have all made mistakes, we have all been swayed by peer pressure, we have all been conditioned into certain training styles and methods, we have all done things unquestioningly and followed the crowd.

I hope these sketches will help put all that in the past so we can learn from them and move forward with a more positive and open mindset, accepting that we can learn to do things better, kindlier, and most importantly, from a place of love with our animals.

They have so many lessons to teach us, all we must be, is ready and open enough to let them in.

I have always been 'sensitive' and have been bullied and criticised for it during my lifetime. Little did I know this sensitivity would turn out to be my 'superpower!'

I am very privileged to have been trained and mentored on my Animal Communication journey by the incredibly gifted Richard Geldard Animal Communicator and Medium.

I embarked on his courses carrying Baggage more than Ego, a lack of confidence, and no real-life satisfaction. I had yearned to find 'my thing', getting lots of qualifications in horsey things that didn't really fill that hole, and never really feeling 'complete'.

That first course, I walked into a room full of strangers but had never felt so much at home, and I have been working as an Animal Communicator ever since.

Mostly, the universe has me helping horses — which is an absolute joy, as I have been horse mad since I was little and have been on my own horsemanship journey for my entire existence.

This really is my calling in life, and I feel so privileged and humbled to have helped hundreds of horses and their humans already, and in turn, they have also helped me.

These sketches are all from animals I have communicated with, or who I have been in the presence of, who have imparted these messages to me, or through me to their humans.

The truth can be painful — but without Ego, it becomes the starting point for a beautiful relationship.

None of this would've been possible without all the darling animals I have shared my life with, from family pets, to those who became the centre of my universe as I got older and into adulthood.

We always had dogs — Red Setters, Springer Spaniels, Black Labradors, a Golden Retriever — cats, rabbits, goats and, of course, horses. Childhood memories as far back as I can remember involved being bundled up in a snow suit, spending what seemed like hours in the clapped-out old car that was the 'tack-mobile', and playing in the old pig houses on the farm where mum kept her horse, a stone's throw from our house. Many, many happy years of a feral 1970's childhood, which left horses in my blood!

My first pony was Tommy Tucker, he was the real start of it all for me. I remember sliding down his neck many times as we came to abrupt snack stops! Gymkhanas — lots of gymkhanas — and one fancy dress class where I was Miss Muffet and Tommy had a spectacular spider costume!

My teenage years were spent helping at a local event yard. The learning curve was steep, and although I learned so much from the discipline and high standards, even then I couldn't help feeling sorry for the horses who seemed to spend most of their time in the stables.

I also saw things that even back then I wasn't comfortable with but were deemed normal. Now I understand the implications for the poor horses, and shudder at the thought of some of the things I'd been taught and that were just thought of as acceptable, and probably still are in some places.

I had a break from horses when I discovered boys, but not for long. Horses were in my blood, but I knew my enthusiasm outweighed my riding ability even then. I joined the local riding club, eventually gaining a place on the committee, which I was so proud of! Finally, I had been recognised as knowing enough about horses to be accepted into that arena.

I did my British Horse Society exams, a must to be 'accepted' into mainstream equestrianism at the time.

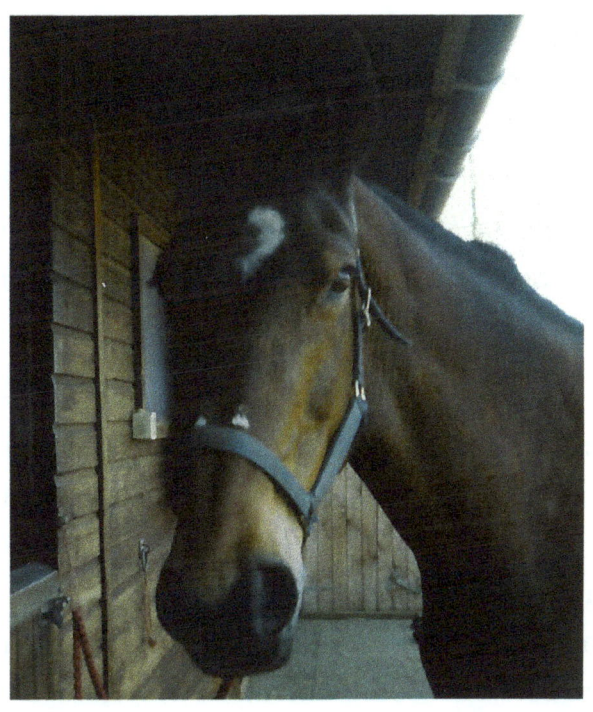

My first horse as an adult was an Irish Draught cross Thoroughbred called Milly, a school mistress who was my first real adult love. I shared her for a while, then one Christmas morning there was a big red bow around her stable door — mum had secretly arranged a sole permanent loan! To this day, that was the best Christmas present ever. Milly and I had many happy years together. We enjoyed hacking, but we did take part in some local shows, and she looked after me so much when I dared to dip my toe into cross country jumping. During this time, I had started to hear voices in my head. It was only in moments of danger when we were out hacking. A few times, I'd heard, 'Be careful — car,' when approaching a blind bend, and sure enough, each time a car had been racing around the corner towards us, going much faster than the bend really allowed. I didn't really question it, and I didn't really explore it either. I did go on to explore what the world now calls 'Natural

Horsemanship'; an ethos of training that is a philosophy for working with horses based on their natural instincts and communication methods. I'd already got the more traditional horse care certificates, but Monty Roberts was a name coming through as a horse whisperer, so having heard the voices, I wondered if I too could do this. But I needed to understand and learn more.

I went to Witney, Oxfordshire, and did a course with Kelly Marks under the Intelligent Horsemanship branding, and went to clinics where Monty was demonstrating how to load 'difficult' horses and 'Join Up'; a phrase Monty uses to describe that moment when the horse decides that it is better to be with the person than to go away. The method is based on non-violent techniques to establish communication with horses.

I was mesmerised and still young enough to be indoctrinated into something on face value. I didn't question it; I simply embraced it, and 'Join Up' became my new favourite thing!

Devastatingly, Milly got colic and although she survived the surgery, it reoccurred the following year; so that was also my first real experience of equine heartbreak.

As empty stable syndrome set in, I felt totally lost. Having a horse isn't just a hobby, it's a lifestyle. My whole routine had changed overnight, and I didn't know what to do with myself.

This led me to Tizzy. A very talented showjumper, she was more Thoroughbred-like than Milly, but she was in my price range, was local, and even when the vet told me she had a heart murmur, I didn't care. All I saw was a horse who needed a good home and even if I couldn't jump very well, I was more than qualified on the good home front; after all, I'd got all my qualifications. What could go wrong?

Well, what went wrong, and very quickly, was this issue of my enthusiasm outweighing my riding ability. Tizzy was much more sensitive than darling Milly had ever been; much more athletic and sharper with her reactions. She didn't ever do anything extreme, and I never fell off her, but she was just too much for me. I felt like I was driving a sports car when I really needed a Fiesta! I ended up getting other people to ride her as I was too scared, but that couldn't be sustained for either of us.

Even then I realised I wasn't being fair to Tizzy, so I visited a local dealer's yard with my horsey Aunty Mary to try and find something more suitable for me.

They brought out lots of horses for me to try but I didn't want to get on any of them. None of them did anything, but I could tell just from their energy that they were still too much for me. This was time to be brutally honest with myself about my truth, as it just wasn't fair on the horses to be unrealistic about my ability and my expectations. Another lesson learned.

In a last ditch attempt I heard an exasperated call of, 'Go and bring out Skippy.' From around the back of the stables, they dragged out this big, orange chunk with a white face, who plodded into the yard.

Instantly I fell in love. I think Aunty Mary sensed it too!

I was impatient for them to tack her up and I hopped on in the riding arena. I trotted around and even managed to jump a little cross pole!

I was smitten, but having had a very strict list from my mum and instructor of 'no mares and no chestnuts', I thought I'd better get them both to come and see Skippy. I knew she was the one but I was terrified of making another wrong decision, and understood my heart tended to overrule my head. I had lost all confidence in myself, my ability, and my decision making, so I went back the next day with them both. As we came into the yard, my heart filled with love when I saw Skippy's big ginger-and-white face over the stable door. I went straight to her and as she rested her head on my shoulder I put my arms around her neck, and we stayed like that for what seemed like ages. An Irish Draught cross Connemara, she was the perfect combination.

Oh, how I love the Irish horse's reliable temperament! When I heard, 'Well, that's a done job then isn't it?' come from behind me, I knew Skippy had won them over too. Tizzy was taken in part exchange and was found a more suitable home with a more capable rider. In that moment I changed Skippy's name to Saving Grace, as that was what she was to me. We went on to have many, many happy years together, doing everything from showing, jumping, dressage, and cross country. Grace was my happy place, so when a field injury meant an end to her ridden career, I found her a lovely home where she went on to be a mum.

While Grace was recovering from her injury, I heard about a young horse who needed a home – free of charge!

This mare was three years old, was living at a polo yard, but was considered too big, so had been left to her own devices and just needed to be moved on.

Heart-over-head mentality kicked in when she was offered to me, but this time, I took a few days to think it over and be realistic about it, for both our sakes.

This mare didn't have even have a name and her behaviour suggested she hadn't had the best of treatment, so I knew I'd have my work cut out. I was told she was 'unbroken' but had experienced a saddle on her back. I knew what that meant in real terms.

So off I went to collect 'Melody', a Thoroughbred cross, with slight apprehension, but also a renewed confidence that Grace had given me. I was adamant this wouldn't be a repeat of Tizzy; after all, I had many years of experience and knowledge now.

My instructor helped me 'start' Melody under saddle and although she was a big girl, she took it all in her stride, and we were soon long-reining her around the roads and introducing her to light ridden work.

It was around this time everything changed for me; with my horsemanship, with my understanding, with all my knowledge, with everything I had spent my life so far learning about horses. It all just turned on its head.

It soon became apparent that traditional training methods didn't suit Melody. I would often see the whites of her eyes and could tell she was fearful. This was also when I got my first real taste of being laughed at and belittled for trying something different with horses.

One occasion, I was at the yard and was showing another livery what I'd learned about the 'Join Up' process. I was demonstrating how to use body language to invite the horse to come towards me, rather than the traditional method of just marching up to them, only to be shot down by the yard owner for not just getting on with it and catching the pony. This lady barged through the area we were working in, marched up to the pony, and caught hold of her headcollar, totally disregarding what we were trying to achieve.

In that moment I realised that human Ego and perceived power trumped anything else in normal horse land. Well, if that was 'normal', I didn't want to be a part of it anymore. It was like I was seeing things clearly for the first time — and I didn't like the view!

She just didn't get it, and as she barged through our activity with no respect for any of us, my whole world changed even more.

In that moment, tears rolling down my cheeks from humiliation and frustration, I knew I needed to forge my own way; a way that didn't depend on who was 'holding the football'.

This is when I discovered the Parelli phenomenon. American horseman Pat Parelli and his wife Linda had marketed a Natural Horsemanship programme where learning started on the ground and slowly progressed into the ridden work.

This programme saved us at the time. It took us to camps, clinics, on forest rides and a beach holiday, and eventually to dressage tests, over the six years we were together.

Melody will always hold a very special place in my heart because she needed me to become something more to help her, and I needed her to help me get there.

This photo was taken on a girl's holiday in Norfolk, where I was able to go out in my dressing gown for breakfast duties and where we could spend undemanding time together; something that I have gone on to learn, from all the horses I connect with, is so very important to them.

I went on to became acutely aware of all the different techniques, all the variations of similar ideology; all very well intentioned, but still, none of them really fitted with what I felt the horses wanted and needed. I found I took little bits from all the different techniques, as horses are not a one-size-fits-all commodity. They are as individual as we are, yet the training methods were all so rigid.

Having been blown away with 'Join Up', I went on to question it. Forcing a horse to go around and around in a small, enclosed space with high walls, with no escape if the pressure gets too much, leaving no choice but to submit, suddenly didn't seem like much of a partnership to me.

Where was their choice? How can they express how it's feeling *for them*?

I felt the same way with the pressure phases from the Parelli programme. The finished product was very impressive on the face of it; horses backing up at the slightest finger wiggle and working at 'liberty' with no physical restraints, just the partnership between the horse and their human.

But, having lived and breathed these techniques for many years myself, I now felt that they were still force-based.

The idea of horse training being called 'liberty' now makes me sad. Liberty meaning freedom. Liberty is just the continuation of the online training, with the halter removed to give the illusion that the horse is doing things willingly and as a partnership, because — look, they are doing it without any restraints!

What isn't presented to the astounded onlookers are the hours of conditioned, repetitive training, involving punishment mixed in with negative and positive reinforcement, to get to that point.

I even went to the McGinnis Meadows Ranch in Montana to learn the American horseman Buck Brannaman's techniques in my quest to understand the Natural Horsemanship phenomenon.

This was an amazing experience, and I learnt how to lope, rope, and lasso anything that moved!

Bucks' ways are more Vaquero, more in line with the traditional cowboy and herdsman methods of horsemanship, rather than the more commercial Parelli methods. I learnt a lot about 'soft feel' and saw, even though the horses were in work on the ranch, they were treated well, and during their down time were able to live as horses, together in wide open space, and as naturally as possible in a domestic setting.

But even though none of these methods solely ticked my boxes, I had enough information and confidence from all these techniques, and from my equine psychology and horse behaviour studies, to move completely away from BHS methods and just do what I felt was right for me personally with my horses.

All these adventures and all this learning doesn't come without mistakes. I've made plenty in my 50 years with horses, but what I have also done is learnt from them, and learnt to listen to the horses.

By listening, I don't mean I just watch their body language and see how they physically respond. I mean I feel them, I connect with them, and I have become their voice. And, as I reflect on my own life with horses and the evolution and growth with all my experiences, I've realised there is a big difference between being an Equestrian and a Horseman. The examples I see around me on a daily basis only serve to reinforce this sentiment, and I firmly believe you can't be both. The horses are our greatest teachers, and the day we put their needs above our own is the day we change teams. Horsemanship is not about the riding.

Melody was the most challenging horse I have ever had. Looking back, I am so grateful she came into my life, as the lessons she taught me have gone on to help me with the other horses that later joined my family. After Melody came Alfie, a feisty little Welsh Cob who, although he was a worrier, was

a confident little hacking pony; unfortunately, a slip in the field resulted in a leg injury that couldn't be treated, so another tearful goodbye had to be said. And then came Anthony. My heart horse.

Oh, my goodness I had never known love like this. A very gangly four-year-old, recently over from Ireland, big and fluffy, with an enormous head he hadn't grown into yet. As soon as I sat on him, that was it. I was in love, again, but I went and tried some other horses, just to make sure. I had learnt my lesson about empty stable syndrome and letting my heart rule my head. But no-one else compared to Anthony.

Anthony was my life, a true gentleman. I was so in love with him. I'd loved all my other horses of course, but this was something else, something that got hold of me and engulfed me. I told everyone that if he'd been human, I'd have married him!

Again, we had a long and happy time together, mainly dressage-related as, now I was getting older, I felt safer with all four feet on the ground, but we did do some cross country and some showjumping.

I never really enjoyed jumping but it was something I got swept along with, as I felt Anthony enjoyed it.

Anthony's passing years later was incredibly traumatic and it's something I still find hard to talk about, even now.

My friend Louise and I were off to the forest. On the way there, the lorry started rocking quite badly and when we pulled over to check them, Anthony was trying to lie down! Our first thoughts were colic and luckily, we were close to a vet, so we diverted there and thankfully got him seen to. Having thrown himself on the ground in the carpark, the vets were amazing and helped me get him inside so they could treat him. After giving him lots of pain relief, they told us to get him home as soon as possible.

On the way home, the rocking started again. Having been told not to stop under any circumstances, we felt we had to try to get home, but it got so bad we made the decision to pull over. We unloaded both horses on the side of the road, and Anthony just threw himself on the grass verge. Thankfully, the vet arrived as soon as he could and gave him some more pain relief to finish the five-minute journey home. On re-exam when we got back, the vet told me we needed to get him to Newmarket ASAP, a good 45-minute journey.

How on earth was I going to do that when he couldn't even manage a 20-minute journey back from the local vets?

This is when my guardian angels stepped in, in the form of livery friends. Sandra jumped in to drive, as I was hysterical at this point, and Louise sorted Prince out and came with us, driving separately so she could lead the way.

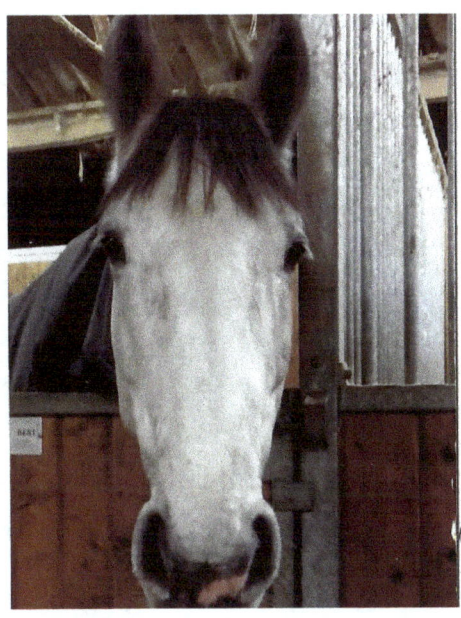

The lorry was quiet most of the way there but when it started rocking again, I was beside myself. Thankfully, Sandra kept calm and just drove.

When we arrived at Newmarket, it became a bit of a blur; lots of people were waiting for us and just took over everything.

After a while of Anthony having tests, we were ushered into a private side room. I knew what this meant.

There was nothing they could do.

I went in with Anthony to say my goodbyes and sobbed into his mane for what seemed like a lifetime. As the pain relief wore off and he started kicking his tummy again, enough was enough.

Unconditional love, but oh what a price to pay to learn about it. I was utterly heartbroken and my whole life had been turned upside down again.

'No more horses,' I said.

Covid hit and along with it, empty stable syndrome got me again. I was browsing local horse adverts online and saw a horse for sale in the next village. Happy hacker, lots of potential, ex racehorse, loving home only.

Perfect!

Graham was duly installed after a test ride where we had walked and trotted, and he had behaved impeccably.

After the test ride, I saw an empty look in Grahams eyes, but I ignored it and thought, yet again, that my love would conquer all.

The next six months were life-changing once more. Melody had set me up well for the future and Anthony had allowed me to expand on all those training methods.

What I hadn't been prepared for was the emotional trauma that Graham was carrying.

Graham was so guarded and emotionally shut down that in those six months, instead of our relationship blossoming, it just got worse. It was a complete mismatch. Not from a place of Ego, but from a place of hope, I felt I could get through the protective walls Graham had built around himself.

But I realised that he was finding my need for his emotional availability incredibly overwhelming, and he just wasn't ready.

Another lesson learnt, and Graham is now incredibly happy with a teenage girl who is doing everything Pony Club with him. I didn't see this as failure but as a learning curve. I realised I was reliving the Tizzy experience through my drive to emotionally connect with a horse again. What Graham taught me is that some horse-human relationships don't work, just like our own relationships with each other. And that's ok; recognising it and doing what's best for both parties is the important thing.

So now we come to Orion, who is exactly how Anthony would've been, if he'd made old bones.

Orion is my soul horse.

He is a big, solid, 17hh, grey, Irish Draught, who I nicknamed the BFG. I thought everyone would know it was Big Friendly Giant, but I was asked if it was Big Fat Grey – Rude!

I saw Orion in an advert and just from that photo, something in me knew we were destined to be together. I immediately decided to go and see him. It was absolute love at first sight, and I was under very strict instructions this time to at least canter him and not to decide on the spot.

I visited and rode Orion several times, insisting I spent time grooming him, getting him in from the field, and just spending time with him. There was nothing that was going to stop me having this magnificent horse, but I went through the motions to keep my family happy. They had a point after all, as my track record wasn't great at horse shopping, let's face it!

Orion has been and still is my Guru, my Yoda. He keeps me grounded and reminds me to stay in the moment. He has taught me how to stay in the present when my mind wanders. He has helped me rummage around in my head and has shown me how to filter the images and feelings, helping me interpret their meanings. He has shown me such patience and wisdom, possible only from an old soul, a grand and higher knowledge far beyond his physical years.

He is my Aslan, guiding my passage through the different realms of existence and energy fields.

He is my Dumbledore, and has shown me how to navigate from the life of a Muggle to being a Spiritual vessel.

I love and trust him with every fibre of my physical and emotional being. He encourages me to learn and guides me unconditionally to help the animals.

He doesn't see me as his keeper; he sees me as his protégé, his partner, his legacy to the Animal Kingdom, trusting me with his teachings to reach out to all species, to share the beauty of the unconditional love we share.

From him I tread this path... For him I will never stop.

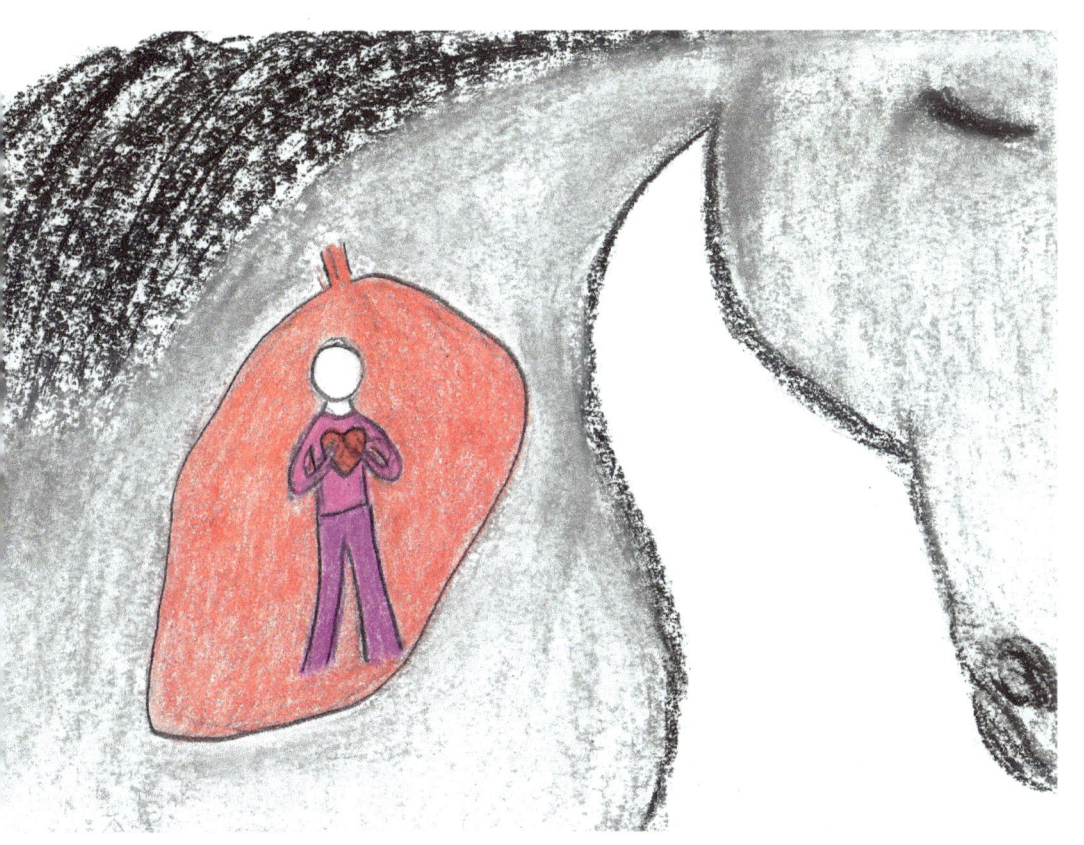

Colours and Symbols

Everyone sees and feels colours differently, based on their own individual experiences and memories. Whilst there will likely be similarities, different shades can mean quite different things to different people.

It is like hearing that special song that takes us back to an event or place and the memories come flooding back, happy or sad, nostalgia filling us with clear images of those times. The colours from those memories are a crucial part of the emotions we are reminded of, and that's how animals use colour when they communicate. They send colours to our mind's eye — knowing what they mean to us individually — to explain how they are feeling.

Take navy blue for example; to me this means structure, rules, straight lines, serious. Why? Well, my school uniform was navy blue, my work uniform is navy blue, so work to me is all those things. Whereas a lighter shade of blue to me is the ocean, the sky on a summer's day; it is the rooftops in Santorini, which leaves me feeling calm, rested, at peace, and relaxed.

Green for me is all about nature, Mother Earth, trees, the forest, and open fields, which feels like being grounded and safe.

Yellow is a beautiful beach, the sand, sunflowers, or a summer's day, which feels like happiness and joy to be alive!

I encourage you to think about which situations in your life you could associate colours with, and to go ahead and explore how you feel with those memories.

I'll leave you a space here to jot down any immediate thoughts you may have about what colours mean to you personally. You can of course keep updating it, as you go along and make new discoveries about the colours in your life.

Have you got a favourite item of clothing? What colour is it? How do you feel when you're wearing it?

These sketches are all genuine real-life situations I have come across, and I have used colour and symbolism to represent the emotions I felt — and how the animal told me they felt — at the time.

I will explain the sketches, some in more detail than others, and some I will leave for you to interpret how they make you feel.

I encourage you to feel the colours, notice the stances, the facial expressions, the tiny subtleties that may usually go unnoticed. I have left you space to write all your thoughts down and to even draw anything you feel inspired to yourself.

To get us started, I wanted to use the first picture to explain my thought process and the communication involved. It is a very personal story and tells how my logo was born, and it signifies the start of my communication journey. This is also why it is the first sketch – no coincidences here!

Ginny Horton

Animal Communicator

Firstly, however, I need to give you a glossary of what some of the reoccurring symbolisms mean.

The rider dressed in formal, hunting and old-fashioned attire represents traditional ideas and equestrianism.

The brown sacks in varying sizes represent Ego and Baggage. The fuller the sack, the bigger the Ego and heavier the Baggage.

Red areas are symbolic of pain — physical or emotional, or both.

Yellow areas symbolise happiness.

Orange is happiness, warmth, and vibrancy.

Black clouds are for negativity, the shades darkening or lightening depending on the depth of despair.

The black spirals are showing anger, shouting, and general bad vibes!

Grey is negative energy or confusion.

Pink clouds are for love; indeed, anything pink will be symbolic of a soft and caring energy from love.

Purple is for healing.

Spirit is symbolised by the angel.

Green is for nature and being grounded, as are the trees.

A red, broken heart is plain to see, as is the red heart full of love.

A light blue in the throat represents the throat chakra, open for communication.

Sunsets depict not only the end of the day but also the end of a situation, or indeed life.

Sunrises are the beginning of a new day, the start of something new and positive.

The differing shades of blue in the skies are for the changing moods, light blue being a summers day; whilst as the sky gets darker, so does the emotion.

The stances of the people should speak for themselves. The main theme is pink for love and purple for healing.

The owl is wisdom and knowledge.

The Robin also features, as he is often seen as a symbol of hope and connection to the Spirit world. He is also representative of not being afraid to be yourself, even if it means standing alone.

The sun rises on a new day, a dawn of a new phase, an awakening. Peace. As I sit in the meadow of wildflowers with Orion, it represents personal freedom, beauty, longevity, infinity. Orion is standing over me protectively as I sit in the field with him. His neck is very long to give complete protection; to surround me, to 'have my back'. I'm in purple for healing; my throat chakra is blue for open communication. My heart is red for love and passion, and I'm mentally sending him the pink cloud of love to his heart from mine. He responds by sending me a pink cloud of love back, but with two golden padlocks, locked together. The padlocks represent us, locked together. To get to this point I had a mountain to climb, and Baggage and Ego to get rid of. I'm wearing purple as dumping all this has helped heal old wounds. The owl watches over us, all-seeing, all-knowing. A powerful Spirit animal guide I didn't know I had until I started to write this book!

All the wild animals are watching on, knowing the truth: that humans can communicate with them, if they want to. They are still, calm, and peaceful, no matter their size, knowing their voices can be heard.

The unhappy dog, the starving horse, the dog chained up, their voices will be heard. The fox, badger, the stag, their voices will be heard. The rabbits, the mice, the birds... Theirs too!

This is how my logo came to be; it was sent directly from Orion to me in response to me sending him so much love in a pink cloud.

A tree, the tree of life. Its roots separating my life into four sections. The last two sections are the most grounded. Top left the journey starts. I'm bullied at school and mocked growing up for my sensitivity. The dark blue surrounding me represents my school years, as that was the colour of my uniform, and for me means structure, routine, a controlled environment. A tinge of black to show it wasn't always a happy time. The feeling of never fitting in. And the book represents immersing myself in my studies — both at school, and something I continued to do throughout my adult life. So many courses, so many certificates, searching for 'the thing' I would connect and fit in with.

I carried a lot of Baggage, and it felt like climbing a mountain: to try and improve, to feel good enough, to fit in. I'm kneeling, dressed in grey, at the bottom of the mountain to represent confusion. As I climb the mountain to get rid of my Baggage, I'm wearing a light green. This symbolises a grounding and a lightness in understanding, and accepting this Baggage needs to go!

Further up the mountain, I'm standing to show increased confidence. I have my Ego in a bag that also needs to go! I'm wearing bright pink to represent love, life, vitality, and an understanding of where my journey is now headed.

The pit at the top is where my bags are going! I am also racing up the mountain on my darling Saving Grace. This is for all the good times, pleasure rides, and cross country we did together. It also symbolises the race to get to the top of the mountain, in a hurry to get rid of the bags.

Bottom left, I then go to 'Natural Horsemanship'; something that helped me with my gorgeous Melody. There was constant judgement around her and what we were doing. This is shown by the three huntsmen symbolising 'traditional' ideas about horsemanship. This new world was no different. People still sat in judgement but hid behind cowboy hats. I am happier here than before. I'm dressed in blue this time, because I feel more structured with my horsemanship, and pink for love and being able to show it with the horses. The grass is a dark green with the huntsmen, getting lighter towards the Natural Horsemanship. There are some flowers to represent 'the grass is greener on the other side' and a fresh start. Even though it still wasn't quite where I needed to be, it was better than before.

Top right, I'm then riding my darling heart horse Anthony. We did a lot of dressage. His legs are deliberately in the wrong position for trot – to symbolise that although we did it, and it looked quite nice, it didn't feel 'true'. We are riding on a stony path to represent the rocky road we went through in life together.

I'm also sitting at the base of the tree, in despair. Both for feeling like I'm doing something I don't really enjoy because I feel I should be doing it, and for the tragic way Anthony went home.

The tree trunk is open because I just want to crawl in and hide, yet the trunk supports me. The leaves are falling from the tree, representing Autumn, and this phase ending. The owl watches over me, ever-knowing.

This is where Orion, my soul horse comes in. The sun is rising on a new phase and gets brighter each day. I'm in purple for healing and my throat chakra is blue for open communication.

I'm at peace and Spirit is with us.

The pink cloud of love with the padlocks has been born and so has the next phase of my life.

This is in acknowledgement of Tommy, Milly, Tizzy, Grace, Melody, Alfie, Anthony, Graham, and Orion. All of whom have taught me so much and helped me get to where I am now.

Love, Light, and Blessings

Before I had even finished the first day at Richard Geldard's Animal Communication course, I was booked onto the Advanced course. I had taken the first step into Narnia and I was fully invested, both feet ready to dive right in. Both courses were completely life-changing in so many ways, and this led me to spending the next six months honing my craft doing photo communications, to raise money for the Brooke Equine Charity. I advertised on local social media, and much to my delight was inundated! I know everyone loves a 'freebie', but this gave me an idea about how interested people were in having this done. I spent hours and hours doing photo communications, learning what signs the animals were using to communicate, which images they used to get their message across, what the feelings they gave me meant, and how to deliver this information, which isn't always pretty.

I soon learnt not to 'flower' things up, not to soften the blow, and not to try and pander to human Ego when passing on their information.

In a couple of cases early on, in my well-intentioned attempts to be diplomatic, to try and make the information a better pill to swallow, I inadvertently lost the imputes of the message and accidently changed the meaning slightly — and therefore changed the impact and subsequent actions.

I kicked myself so hard when I realised what I'd done as I hadn't been true to the animal, I had put the humans' emotions first; I hadn't been a true voice for the animal.

Like many things, this is a learning experience, and I have never done it again. In fact, I make a point of telling my clients that I don't dress things up to save their feelings, that I say it how it is given to me; after all, it's not my information, and unfortunately if the truth hurts, well, at least that's a starting point for change.

After six months I felt I was ready to tentatively venture out for my first communication visit. I got my insurance and just wanted to gain as much experience as I could, so when a friend mentioned she may know a potential horse and human who might like to try, I jumped at the chance.

This was an incredibly pivotal day for me. I had gone on the courses for my own learning and to help me communicate with Orion, and yet, here I was, heading out to help other relationships. Was this what was in store for me all along?

This was an incredibly emotional communication, and any nerves I may have felt vanished the moment I connected with this beautiful horse. The information and tears flowed but the communication was followed by such an overwhelming sense of peace.

Ah, so, this is why I'm here... Finally.

I wouldn't accept any payment for this session as I didn't feel I'd completed my self-inflicted probation yet. It never was and never will be about the money, and I felt thoroughly overwhelmed with what had just happened.

A few days later I received a 'thank you' from this lovely lady and a voucher for a drum massage. 'That sounds nice,' I thought, but had no idea what on earth it was.

I started doing more and more communication visits and went on a one-to-one training day with Richard to further enhance what I'd already learnt. I wanted to spend time with him to soak up his fabulous energy and positivity, to gain insight and a deeper understanding of what I'd developed so far.

Months passed by and I'd forgotten about the drum massage, but after feeling like I needed some self-love, I booked in, still having no clue about what it was I was going to experience.

After a lovely welcoming chat with Michelle, I laid on her couch with my eyes shut and just listened. There was some beautiful relaxing music playing and then I heard the drumming start. I kept my eyes shut the whole time and I could feel her moving up and down alongside my body at the side of the couch, all the time drumming over my body. I could feel the beat, the vibrations, and it seemed to get incredibly loud at intervals. She seemed to spend a long time at my feet.

My mind started to make images that I couldn't make sense of, yet they seemed so real. I could see a castle-like building; it was dark and dingy, there were little ninja-like creatures scaling the walls, there was a dark tornado heading our way, and as the drumming got louder and louder, the tornado got closer and closer, and then the ninjas started falling away from the castle wall.

Just as the drumming crescendo finished, everything disappeared. But it was fixed into my memory, even today.

I told Michelle all about it and she was surprised I had remembered anything, let alone in such detail. She said people normally just fell asleep! Michelle said that what I'd seen was negative energy being cleansed.

The image was so vivid that I felt I needed to release it somehow, as it was just taking up all my brain space. So, when I got home, I attempted to draw what I'd seen in my mind's eye.

I have never been any good at art, I have never been able to draw, but I did manage to replicate a rough idea of what I'd seen and sent it to Michelle, who said I needed to do a burning ceremony with it, to release the negative energy.

I did this and didn't think any more of it. But out of nowhere later that day, I had an unusual urge, an itch that couldn't be scratched; an absolute need to get everything out of my head and put it on paper. I needed to clear some more head space!

The meditations from Richard's courses were my first works! The images from my meditations came to life before my eyes and although childlike in structure, portrayed my safe place and my Spirit guides in a way that at least they were recognisable.

The next few days and weeks turned into a complete blur. I was drawing everything and anything.

Richard was working on his Whispers in the Wind single at the time and when I sent him my interpretation of him with the animals, his next request took me by complete surprise.

Richard asked me to design the CD cover!!!! Gulp! I thought he was joking as I felt my sketches were so childish. He wasn't joking, quite the opposite in fact. Not only did he want me to design the CD single cover, but he also wanted me to illustrate the video!

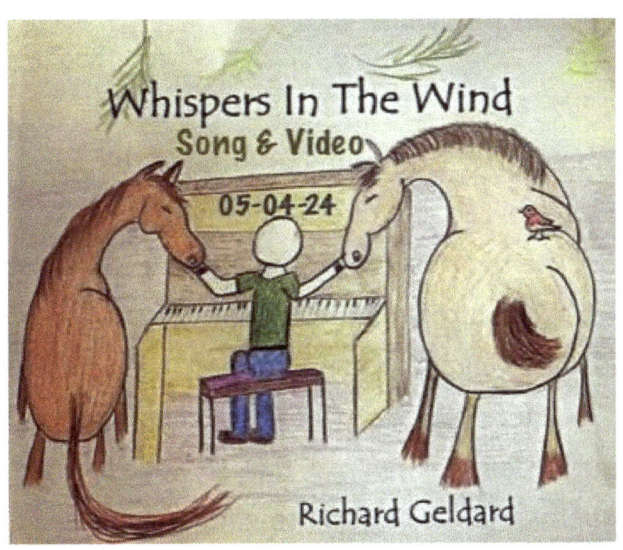

The 'gift' of Animal Communication isn't the communication itself. It's being able to lay ourselves open and bare enough to allow the communication to come though us. Everyone can communicate with animals if they want to, but it involves moving past our Ego and Baggage, trusting in Spirit and in ourselves that it will happen. Here, we see the dark blue sky on the left for structure and order, the person has removed their clothes to literally lay themselves bare, the brown sack of Ego and Baggage is ready to be emptied over the side of the cliff, the wooden bridge is rickety over the deep ravine, but on the other side, Spirit is waiting with open wings. Trust is needed to cross that rickety bridge, symbolising that it's all about trust. The grass is greener with Spirit and the sky is a lighter shade of blue, peaceful and calming. Spirit is always shown in Angel form, and purple for healing.

Self-refection and 'looking in the mirror' are some of the hardest things we'll ever do. To be honest about our true emotions and behaviours isn't always pretty, but it's something that must be done to get rid of Ego and Baggage. Here, we see the metaphorical mirror, with Ego and Baggage sacks at the top with the prized 1st place red rosette, reflecting on what drives the need to win them. Conditioned behaviour, belief systems, peer pressure, confidence, pride, and traditions all contribute to Ego and Baggage, and it's something we all have, even with a little 'e'. Ego doesn't have to manifest itself in sports cars and designer clothes, it can be as simple as competing when we know our horse isn't 'quite right', riding our horse regardless if they've shown they don't really want to by being 'difficult' to catch; Ego is 'showing them who is boss'. Once we let go of all that Ego and we see the world from their perspective, everything changes.

Meditation is a huge part of Animal Communication. It allows us to concentrate our mind, empty our own thoughts, and create space for the communication channel to be open. Meditation isn't just for the communication. It also allows us to connect with Spirit and take us to meet our Spirit guides, who will help and guide us along the way. Spirit guides come in all forms and species. For me personally, I have an extraordinarily strong connection with Native Americans, a Wolf, and a Monk. Here we see how Spirit guides help open the gateway to communication through meditation. The owl is often present, ever-knowing.

Once all this has been achieved, we are ready to embark on the actual communications. This sketch depicts the celebration of me finding and accepting Spirit into my life unconditionally. The Native Americans are playing shamanic drumbeats, the candles are lit to also show the way through love and light, my guides are there, animals are peeping from behind the trees, ready and waiting to be seen and heard once the ceremony has been completed. I will be walking down the steps to signify the journey to get to this point, and past the graves of Ego and Baggage.

Then, after descending the steps I find my own special meditation place. My personal place for sanctuary and safety is a log cabin in the woods. Here we see me as a Native American following and trusting my Monk, with my Wolf protecting me as I head into the unknown that will soon become my new life.

When I arrive at the log cabin in meditation, there is always a huge fire. On many occasions there is a Native American tribe — my tribe — there, waiting to greet me and celebrate my passing into this part of my life. There are always women with babies, signifying The Mother in Mother Earth and Mother Nature, and the birth and rebirth of Spirituality.

The Tribe itself is also very significant as it is the collective noun for like-minded people, where there is safety and no judgement, only unconditional love for each other, animals, Mother Nature, and Spirit.

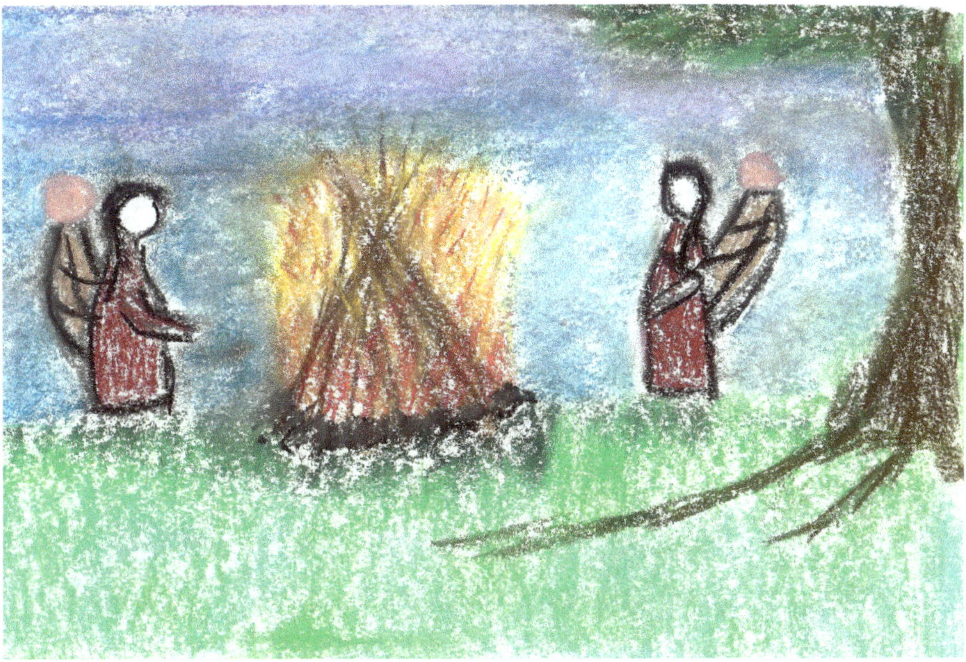

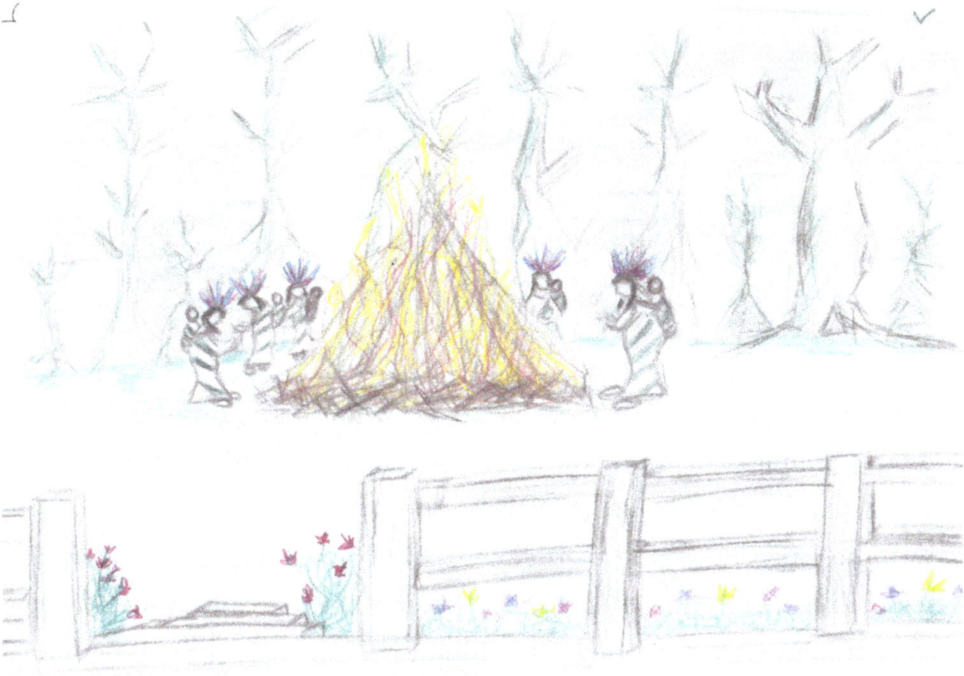

Belonging to a Tribe becomes like being part of a family who you choose yourself. Living with nature, only using what you need, not being greedy, living without judgement of others, and loving everyone and everything around you. Easy to say but not necessarily to do. Here the Natives live with the buffalo but only hunt them when they need to eat. They thank Mother Nature for the sustenance and give thanks to the buffalo for its sacrifice for their survival.

Through meditation, letting go of Ego and Baggage, self-reflection, and the power of intention, we learn to make the Spiritual connection with the animals. We learn how to protect ourselves from negativity and to keep our circle of love and light surrounding us. We learn to visualise the connection with Spirit, and how to allow the energy and information to flow freely through us from unconditional love.

We must also remember to close the connection in the same way after each communication.

When we start on the communication journey we must learn to block out the outside noise and any negativity that may be present. We must trust our own heart and gut instincts. The clouds on the left are various shades of grey to symbolise confusion.

The horses' hearts are clear to see and feel, all so honest, so truthful, and are all coming from unconditional love.

Spirit sits with each horse, offering communication and healing if we're open to it. I have my hands over my ears, trying to physically block out all the negativity. I am dressed in blue trousers to symbolise routine and structure that is familiar, the pink top for love. My heart is big and red for passion and life, the heart in my stomach is purple for healing. My throat chakra is also purple, as the communication will heal the confusion, and listening to our own heart and gut instinct will help heal them all.

If unconditional love is our true intention, we must trust our gut instincts; they will navigate us through the negativity, and healing will come.

Spirit will help guide us if we trust. And this doesn't always mean taking a deep breath and doing it anyway, as no species learns through fear or pain.

When our horses get physically between us and another person while we're talking, or when an energy shift occurs around us, it can be a sign of protection and a reminder to maintain our boundaries. This manoeuvre comes from so much love from the horses and is so subtle, yet so obvious when we recognise it.

It can be from well-meaning friends or from someone who perhaps isn't having the best day. The tree is grounding and its beautiful flowers are pink and purple for love and healing. The horse's neck and tail sweep around their human protectively, linking up with the tree roots representing security. The flowers just float in the air and fall gently, creating a subtle physical barrier; a loving reminder that creating boundaries doesn't have to be negative, but a part of loving and protecting ourselves, and us and our horse as a partnership together.

The rest is not in colour, as it's the boundary that's the important part of this message.

Listen to them, and trust their advice.

Listening to horses doesn't have to mean being Spiritually connected with them. But, without realising it, we give off emotions and energy based on our expectations. Here we see the rider arrived ready for riding, because that's what they have decided to do, irrespective of if the horse really feels up to it. We also have someone who has come to see their horse, not dressed in riding clothes and without the tack, just checking in to see how the horse is before making that decision. This is also another example of how Equestrians differ from Horsemen.

The weight of expectation lies heavy with some. I know how pressured I have felt in the past when I was busy being a people pleaser or did things begrudgingly. The unspoken words, the air of expectancy and assumption lingering and hanging over my metaphorical head, really put a lot of pressure on me that I didn't realise at the time.

It occurred to me when I was getting dressed that I no longer wear jodhpurs or riding trousers, always wearing jeans when I go to see Orion. A subconscious decision based on my intentions. My intention to arrive at the yard and just see what happened. If I ride, all well and good, if I don't, equally all well and good!

Awareness and power of intention is something that becomes a lifestyle, and because I understand that my emotions and energy towards Orion start with my intention towards us and our activities, even before I get to the yard and getting dressed, I realised I didn't want to put that 'expectation to ride' on him every time I arrived to see him.

Because of this, Orion is a much more willing partner and says 'yes' to riding more readily, because he knows that if he says, 'Not today, thanks,' he is listened to.

By wearing jodhpurs my intention was to ride, and then there is potential pressure for Orion to agree to it even if he doesn't want to, out of conditioning.

I don't want that for him. I want him to be able to be honest and to trust me, to have a choice, and to be confident in making and expressing that choice, which he does.

This is important to me in my work as a communicator; to break down every thought, every intention, and every action to get the deep connection he and I share together.

He's teaching me so much and I am so blessed to have this beautiful, gentle giant teaching me all these wonderful things.

So, let's explore the impact of us having a bad day and taking that energy to the stables with us.

Here we see a flat tyre and dropped shopping under black clouds to symbolize the human having a bad day and negative energy. They are angry, so take it out on fellow humans who don't deserve it and are confused by the reaction. The other people are surrounded by pink and light blue showing their energy is gentle and kind. The person's bad day gets worse as shown by more blackness and the head being completely black. The black energy goes with them as they approach their horse; even their stance feels aggressive. The poor horse, coming from pink unconditional love, is frightened by this negative energy approaching, and then becomes affected themself by the negativity.

A similar situation here. The person has lots of Baggage as they are suffering from heartache, which is making them full of resentment and negativity. They take it out on the poor horse, who doesn't understand why they are being treated this way, which then leaves them also broken-hearted.

Even the ever-knowing owl has his head under his wing, trying to hide from this awful atmosphere. He can fly away if it gets too bad but the poor horse has no escape. This then leads to depression in the horse, resulting in poor condition and an unwillingness to partake in activities with the human. The human then interprets this as the horse being 'difficult' or 'naughty'.

The horse is simply reflecting and mirroring the human's behaviour.

Competing can bring out the worst in us. For some, show nerves result in a quick temper, and the pressure we put ourselves under to do well can overshadow the enjoyment. This can cause us to be less patient than normal, overreact to small things, and can result in the horse reacting to our energy before we've even left the yard.

Over the years of cross-country jumping with Grace, we rarely completed the course without being disqualified, usually within the first three fences! We did lots of course hires in between events to practice but on the day, the pressure always got too much for me! We usually did pairs, and it was only the smaller classes, to encourage Grace to follow and to support her where my nerves let us down. I remember being furious with a friend one day when half way around, Grace wouldn't go into the wooded area; she refused point-blank, and even the fact that her friend was cantering off ahead wouldn't convince her to follow. I was furious — not with Grace, as I understood that the light had changed and she didn't feel secure — but at my friend, who just carried on and completed the course, 'going clear' without us!

Of course I understand her perspective now, but I remember feeling that I'd let Grace down by not supporting her through the trees as a team, to build her confidence for next time; and not understanding that Grace wasn't my friend's responsibility and she wanted to enjoy her jumps!

There is the ugly side of competing, in all disciplines, when the desire to win goes too far, and it's always at the expense of the poor horse.

Even at these local events I used to attend, the amount of frustration I saw being taken out on the horses was disgusting.

The horses who were whipped mercilessly for not wanting to go through the water jump. The rider forgetting that horses have a different depth perception to us because their eyes are further apart, and their instincts were protecting them from water dwelling predators.

This is something I had to work on with Melody. We spent hours and hours in-hand together, me going ahead in wellies, her following at varying speeds until she trusted that I wouldn't lead her into danger. I built her a little paddling obstacle in her field with poles and tarpaulin, so she could just get used to walking through it and gaining her confidence with placing her feet in the water.

Yet again, as humans, there can be an expectation that our horses are able to just switch off thousands of years of instinctual and genetic behaviour.

In order for them to overcome this, there needs to be trust, and that trust only comes from a loving and respectful relationship, not from fear.

Your turn now. What do you see here and how does this picture make you feel?

Please feel free to jot down your interpretation of what you see here, using what you have learnt so far.

Loading is often a situation where Ego takes over, even for the calmest and most patient of people. The perceived pressures of time and being late, the feeling of other people watching and criticising, the feeling of frustration because they went in fine the other day, all add to the tension of the situation.

Loading is also an area where we are likely to be offered 'help' without asking for it, usually in the form of lunge lines, whips, flags, brooms, blindfolds, feed, and pressure halters.

We all know that it's a massive ask for a horse to go into what is essentially a cave on wheels. The travelling is tiring and can be painful if there are medical issues, diagnosed or not, going on. Previous experiences play a major role in how horses feel about travelling, and many have told me that they have been frightened because they can't get their balance because their human drives too fast.

Loading practice should not be done on the day of the outing and any gaps in our relationship will soon become evident during this activity.

This can also be a time for mirror gazing and working on the relationship between us. There are many reasons for loading — vets, competitions, pleasure rides — but each time involves an enormous amount of trust from the horse, which shouldn't be underestimated.

This sketch shows the clock with hearts instead of numbers, to take the time it takes, time is kindness. The person at the top of the ramp is not only physically blocking the way, but their stance and what they are saying is aggressive and negative, affecting the horse directly emotionally.

The ever-present helpers are on standby with the 'helpful' comments and tools. The most important part of this sketch is the horses near hind. The reason for reluctance, the red area is representing pain.

Never underestimate the value of undemanding time together

Horses are not machines. They are living, breathing, loving, sentient beings and souls.

Human expectations and rigid routines that demand perfection from their energy source, will only bring emotional discomfort for the horse.

We need to examine what Baggage we are carrying and what is driving our Ego, and how that is manifesting in behaving this way.

Horses who are living in an energy field of training and competing with no consideration for their emotional wellbeing are suffering, which will then manifest as physical issues in some cases.

How many times when we get fed up, stressed, or are anxious, do we get head- or stomach-ache? How many times do we feel anxious when we are worried or frightened? How many times do we feel sick with nerves?

The horses are not any different.

I don't do touchy-feely!

Our own energy affects our horses

The horse breaker, replacer, and forgetter. So many people use horses as machines and for their own gain - as long as they are useful in some way. I understand horses are expensive to keep but there are people who when their machine is broken, they just go and buy another one. The broken ones are just left with no attention and discarded. The human convinces themselves that the broken ones are put out to retirement and don't really need much attention.

If the horses are simply left in isolated paddocks, hardly groomed, or spent any time with, that is not retirement. Here, top right, we see the introduction of the well. But, if they no longer serve a purpose, the Ego of the human who is fuelled by their own self-absorption and bitterness, and thinks they know best by 'retiring' their horses like this, is actually no better than treating them like rubbish by ignoring their emotional needs. Their future is to eventually end up on the collector's scrap heap, waiting for their eventual fate.

Instead of spending more money on a new horse who will also end up on the scrap heap, perhaps we should reflect and ask ourselves why our horses all keep breaking, and look in the mirror.

Human Ego, Baggage, and Pride all come at a price for the animals.

Being stuck in the bottom of a deep well is how the horses show me they feel forgotten about.

Being in a deep well, amongst the overgrown weeds, with no escape, just waiting for someone to come, with a complete feeling of helplessness.

They often don't have eyes to show they don't feel seen.

Thank you for everything, I'm ready to go home now.

It's not goodbye, it's goodnight darling one

We do this one last thing

From unconditional love, from the bottom of our hearts

We thank you for all the good times and will cherish all
those precious memories

And all the love that we shared

Run free now darling one

You will always be in our heart, and will never be forgotten

Until we meet again...

Being asked to communicate with elderly or very ill animals, to ask if they are ready to go home, is not something I do lightly.

I consider it an honour and privilege to have been their voice in this situation many times now.

It is a massive responsibility for me. I am in no way 'playing God' or influencing any decisions to be made. It is always the animal's information, and we have a long conversation about their physical and emotional pain before I ask that question.

Animals don't see death as the end. Richard Geldard, Animal Communicator and Medium, explains:

'The journey of the animal's soul is no different to ours, it is to experience a life on this earth and to take back their experiences to Spirit. The journey involves experiencing death of the outer body and the continuation of their soul.'

Mostly, by the time I'm called for this conversation, the animal is ready, and it has been a difficult conversation to listen to for their human. However, everyone who has contacted me under these circumstances has always messaged me a few days later to thank me.

They all said that the guilt had been taken away from making the decision, and they had taken great comfort in knowing that it was the right time and their animal had known it was their last act of love for them.

Deborah and Mondi

Deborah contacted me as she was worried about her elderly mare, Mondi. Mondi hadn't been sleeping very well, was struggling physically, and Deborah really just wanted to know if Mondi was ready to go to sleep, which Mondi later confided that she was. During her photo communication Mondi showed me a tree and a stump next to it, saying this was their special place together. Mondi wanted Deborah to go there to remember her after she'd gone. I tried to draw what I'd seen to show Deborah as it was so specific in my mind's eye.

Deborah then sent me a photo of a tree in her garden, with a stump next to it that had been made into a seat. The photo was of them together at this very tree in Deborah's back garden, during Mondi's younger years.

I wrote down the feelings I had while I was doing the sketch for Deborah to keep. This is how *Soul Sketches* came to be and are now being used to help and comfort so many, as Mondi's legacy.

Deborah Carpenter + Mondi 5th Feb'2

To Deborah, Your special place by the tree.

Mondi looking over you, seeing you from her own eyes, knowing you are thinking about her.

A meadow of wild flowers. This is what horses show me for 'freedom'. Full of life and colour and just so beautiful to look at, stretching as far as the eye can see – infinity.

The sun setting, symbolic of the physical time coming to a close. Not the end as it will rise again in the morning; life goes on in so many ways. Flowers planted by the tree in memory of your time together, but also as a constant reminder that there is beauty all around, in the smallest of things.

Deborah's Words

Making the decision to have a beloved horse put to sleep is probably the hardest thing we ever face. I had been agonising and losing sleep over this for quite a long time. I didn't know of Ginny, but something led me in her direction and my instincts told me to contact her.

Ginny did a remote communication (from a photo) with my 31-year-old mare Mondi. It was a clear moonlit early evening, and I observed Mondi from inside my house whilst Ginny was in contact with her. I knew something very special was happening and I felt a strong connection between the three of us. Mondi told Ginny that she was very tired, she knew her body was failing, and she was ready to leave the physical world. Lots of other personal and emotional information was shared and I knew then that Mondi had made the decision for me.

I will be forever grateful that the communication helped me cope with the heart-wrenching 'last act of kindness' I could do for my special mare.

I know she is still with me in Spirit and is guiding me on my horsemanship and Spiritual journey. Ginny, Mondi, and I now have a special bond, and I'm so pleased that Mondi's legacy lives on through Ginny's inspirational drawings.

I now often sit at our special place, the curved bench under the tree, which is the picture Mondi sent to Ginny that night, and that she wanted it to be my special place to remember her. I've since planted a red rose which is named 'Precious Love'. Yes, there are still tears, but more often a tranquil feeling of love, connection, and gratitude.

To quote Ginny from that very special evening 'Mondi says it's not goodbye, it's goodnight.'

I will always be grateful that I found Ginny. When my beloved mare Mondi was nearing the end of her long life, Ginny's communication, with her empathy and honesty, helped us both more than words can express. The equestrian world needs forward-thinking people like Ginny, and now that she is working with open-minded professionals, I have no doubt that many horses near and far will benefit enormously from her gifted communications.

Thank you always, Ginny, from the bottom of my heart. Deborah and Mondi x

Kelly and Sheeny

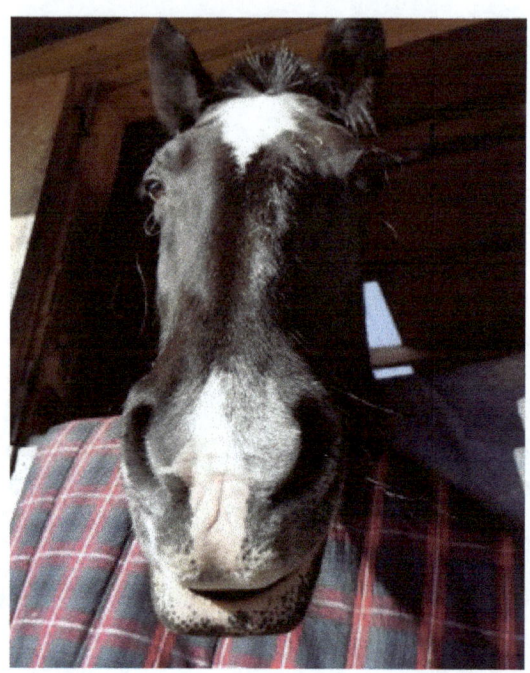

Kelly contacted me for a live video link communication with her beautiful mare, Sheeny.

Sheeny was mature with some health issues, so Kelly wanted to check on her physical and emotional wellbeing.

We chatted over her ailments but during the communication, Sheeny seemed unusually distracted and kept staring off into the distance. Kelly mentioned it a few times but couldn't see anything herself, so we carried on with the communication, and although Sheeny knew she was experiencing some health issues, in that moment, she felt she could cope with them.

The staring out into the distance continued, and although puzzled by it, we weren't worried.

Kelly video called me unexpectedly early the following morning, very distressed. Sheeny had gone down and couldn't get up, and Kelly was waiting for the vet.

At Kelly's request, I connected with darling Sheeny via live video link and passed on her thoughts and emotions as she prepared to say *goodnight, not goodbye* to her devoted mum.

An emotional call for all concerned as you can imagine, but Sheeny was very relaxed, and safe in the knowledge that a horse called Badger, that looked like her but she'd never met, was waiting for her.

Kelly didn't know who Badger was but was sure he'd been there for a few days, as she'd caught glimpses of Sheeny staring off into the next field, mesmerised by something, but not by anything Kelly could see, like she had been doing during our communication the night before.

The following day Kelly sent me a message saying 'My lovely neighbour came round with some flowers and a chat. We bought the house from her, and I told her about Badger. I said you don't know a Badger, do you? She said yes, Badger was a horse who used to pull the canal boats down at the wharf which is at the bottom of the hill!!!! You know how you said he looks like Sheeny....'

Photo Courtesy of Roger Fuller

The badger, symbolic by features, but also possibly by name of a horse in Spirit, not known to Sheeny in the physical. Badger waiting for her in the next field. The gate is open but has mud underneath to symbolise it's been a bit sticky recently. The two purple angels are Spirit, the Robin represents Spirit in the physical, so Sheeny wants you to know that when you see a Robin, she's with you. A massive heart, to represent her personality but also the abundance of love she received and gave while with you. She insisted on not having the traditional rainbow bridge, instead she wanted patches, like a quilted blanket. The red on the bridge by her left ear is the pain she was in. It's behind her now, and the colour purple for healing is the final patch before reaching Badger. The sun sets behind the clouds, but it isn't 'goodbye', it's 'goodnight'. The green grass is for her valiant attempt at eating in the neighbours' field despite her health issues.

Kelly's Words

'My little Robin comes, she followed me on a whole walk with the dogs the other day. My friend lost her elderly pony about a month ago and I was messaging her, she owns her own livery, I told her about Badger and the story. The next morning she checked her CCTV from before she went to bed and saw a badger running right near Hannah, her pony's, paddock. She has never seen a badger on her land before!'

Kelly xx

Aoife and Bubbles

When Eilish asked me to visit her two horses, I wasn't expecting it to be such an emotional morning. I'm used to a few tears here and there, but I hadn't been prepared for the sadness that came to light during our communications that day.

I started with her big boy, Arg, and it soon became very apparent that there was a deep void emotionally between them. When Eilish told me that was exactly how she felt because she was mum to her beautiful daughter Aoife, who was now forever three years old.

Eilish was grieving and had found it incredibly difficult to even go to the stables, and had completely shut down emotionally.

Aoife's pony, Bubbles, was less emotionally vulnerable. In fact, she knew Aoife was with Spirit and as Eilish and I were talking about how she thought about Aoife as a butterfly now, one landed on her shoulder, and she said how a Robin regularly visited.

Aoife is in a pink cloud of love with her beloved pony, Bubbles. Spirit surrounds and protects her while also symbolising her Spiritual existence in the physical as a butterfly and, of course, with the Robin. These messengers have kept her close to mum Eilish.

Eilish's Words

Aoife is forever three. She had her cancer diagnosis for just five days before she passed away, she was misdiagnosed and I was labelled a paranoid mum. She saw a doctor or went to A&E in excess of ten times in the weeks leading up to her passing away. It was very traumatic for all of us, and I can 100% see why Arg had so much anxiety as to what happened, as we were always together before. Aoife's Bubbles charity is named after her and her beloved pony, and is dedicated to her memory, the UK's only registered childhood Germ Cell Cancer Charity. Their aim is to raise awareness and educate as many people as possible, and provide acts of joy to children and their families affected by childhood cancer as a whole. www.aoifesbubbles.co.uk

Karren and Dela

Karren and I have been good friends for some time, and when I started on this journey she was one of the few who openly supported me every step of the way. We used to ride out together regularly, hilariously looking like little and large on our two greys. I nicknamed Dela Thumbelina as she was so tiny in comparison to Orion, and we have many lovely rides to reminisce about — along with some hairy moments, when Dela 'suddenly' and 'out of nowhere' started to buck while out hacking, and on more than one occasion I glimpsed Karren flying through the air. The usual health checks were undertaken; vet, physio, saddle, teeth, feet, everything was investigated.

On a couple of occasions, usually about an hour into a ride, Dela would just plant her feet and refuse to move. Some were quick to label her a 'naughty' pony but on one ride, I could see her tummy muscles so tense that it was obvious it was pain-related. Karren immediately dismounted and walked all the way home.

There was intermittent lameness identified; a shoulder, the back, and hocks. She was on varying periods of investigations, return to work, and rehab, but Karren was convinced she still wasn't right. An MRI scan confirmed that there were issues in several areas.

About this time I had completed my first course with Richard so was eager to practice, and by now Karren was gratefully accepting any help that could point her to do the right thing for Dela, even if it meant her being retired from ridden work. Karren didn't care about the riding, she just wanted Dela to be pain-free.

We did several communications with Dela over the next few months. Karren completely trusted the process and trusted in me, which is a responsibility I took incredibly seriously.

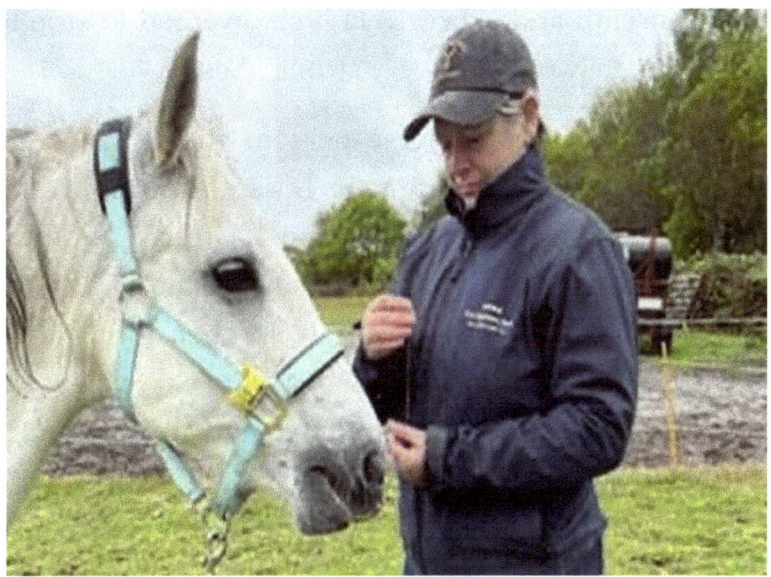

Dela gave me a strange image during one communication. I had learnt to trust these random apparitions, so I mentioned it to Karren hoping she would know what it referred to. What I saw was a monkey's face, like a child's soft toy. Karren couldn't think of what it could be, so we carried on with Dela's questions.

When I got home I searched the internet for what Dela had shown me and this is the face I'd seen.

Photo taken from Amazon

We checked in with Dela regularly and she got the treatment she asked for at each phase, but it soon became clear that the ridden work was aggravating the physical issues. So, Karren took a deep breath and asked Dela if she wanted to stop being ridden and the answer was a very strong, 'Yes.'

Karren is a like-minded horse person; she believes, where possible, horses should be able to be horses and our livery yard didn't have a herd turnout at that time. There were only sole-owner turnout paddocks available, so that was the next quandary for Karren. It was an agonising time for her, trying to work through her needs as a human, and wanting to care for Dela herself and keep her close, or moving her to a retirement herd where she could live out the rest of her days as a horse.

Not an easy decision as a human but when you put unconditional love first, there is no real choice.

Also at this time Karren had welcomed a new grandson into the family, and messaged me when she next went to visit him as she'd been sorting through his toys and her daughter had pointed out his favourite toy!

Before making any final decisions and just to make sure, we asked Dela what she wanted and gave her the choice. She desperately wanted to live in a herd, and this is a classic example of the truth not always being what you want to hear — but then Ego is tested to the core!

After a lot of research and recommendations, a retirement herd was found who had a Dela-sized vacancy and so, after a conversation with her to tell her what was happening, Dela moved in. She made firm friends very quickly, and Karren's frequent visits gave us reassurance that she was happy and settled. I even went with Karren to check with Dela to make doubly sure.

She was as happy as could be and, although was a little more 'roughed off' than she was used to being, had installed herself quite happily into the herd hierarchy.

Karren received regular updates over the next 18 months but one particular message on a Saturday morning, saying that Dela was 'a bit footy', turned into a more urgent message the following day. The vet had been called to Dela; she had life threatening laminitis.

The following weeks were incredibly traumatic for both Dela, as she had to be moved to a stable away from her friends and was in excruciating pain, and for poor Karren who suffered terribly, feeling completely helpless.

When Dela was well enough she was moved to a more suitable rehabilitation yard, where she was watched around the clock and had company around her. An all-too-brief improvement lifted Karren's spirits, although the magnitude of how bad Dela actually was was never far from her mind.

Then, the dreaded phone call came; Dela had taken a turn for the worse and, realistically, unconditional love needed to be called upon once more.

Karren's Words

I was 50 when I had my very first own horse; it was 2017 and my trainer knew of a horse she thought would benefit from a more one-to-one relationship. Derry came into my life, and I fell in love. He was labelled as being 'naughty' and I was told to just give him a whack. It wasn't in my nature to be like that with an animal, especially when we had a real bond. Derry had a very cheeky personality but was so very gentle. I didn't really understand how close the bond was until he collapsed on the yard and wouldn't let anyone near him apart from me, it broke my heart when I had to say goodbye to him!

I didn't intend to have another horse, but empty stable syndrome was in the air. My goddaughter had known a very gentle mare for four years at the stud where she worked, so I went to meet her. She was very sweet and so I brought her home October 2020, and that's where my real learning journey started.

Again, she was labelled as a 'naughty' pony and a 'stubborn mare,' but I'd learnt a lot from Derry, and I tried to learn to listen to her. By the summer of 2021 Dela had intermittent lameness. I had started to ride with Ginny and the BFG and Ginny was starting her journey of communication. With her great knowledge, she helped with Dela's rehabilitation. After every incident, we found something else that wasn't quite right. After all the vets' visits and treatments, Ginny communicated with Dela and the more we listened, the more we learnt that Dela was very good at letting us know when things weren't right.

Ginny supported us every step of the way, not only by communicating with Dela but also helping with her rehab. Her way of helping is truly special, as everything is done from love and kindness. I lost my Mum in September 2022 and in November 2022 I had a long-distance photo communication for Dela with Richard Geldard. Derry came through and the fact that Dela knew of my loss and understood my pain (Richard did not know my Mum had passed) was just amazing. There was one other thing mentioned in my communication that no one could have known about, and in a way that sealed my complete faith in the process of trying to communicate with animals and giving them their voice.

Dela wasn't with me for a long time, but her impact was huge. During the investigations, rehab, and physio, we spent so much time together and looking back, I know she carried me through my darkest days after the loss of my mom. I truly wanted to give her the most natural, happy retirement. I remember the communication with Ginny where we actually asked if she wanted to be ridden and she said no she didn't. I was sobbing, not through the fact that she couldn't be ridden any more but the fact that she was in pain and was suffering and couldn't tolerate a human on her back. Without communication, Dela's story may have been different. Unheard, she may have been pushed into things that caused her even more discomfort.

I just wanted a long, happy retirement in a herd for her, which unfortunately didn't happen. Again, the day we went to try and move her when she was so sick, was a testament to how far we'd come and how she trusted me. Dela knew I would listen to her, she knew we were trying to help her, she was in so much pain, but she eventually got into the box so we could move her to stable her. I went every day; she would lean right into me, she knew we were trying everything to get her better, I know she felt my love. Ginny was so supportive during this time, and was always there and at the end of the phone during this terrible period.

Lost Cat

When I first started doing communications for other people, to get some practice I responded to a 'lost cat' advert in a local village. I didn't know the village or the distraught lady whose beloved cat hadn't been home for a couple of days. Herself and her husband were beside themselves with worry. He was a Bengal, and they had paid a lot of money for him. They had even put a GPS collar on him so they could track his movements and know he was ok. They had found his collar but there was no sign of him. I asked for a photo and set about trying to find him with Spirit's help.

I immediately got a very strong connection with him from the photo. A stunning-looking boy, and a real feeling of a proud and free soul. He was happy to communicate with me and was very clear and adamant with his information.

He told me that he wasn't far away from his humans, in fact

he'd seen them looking for him. He described the thatched cottage with flowers growing up the walls and two golden retriever dogs in next door's garden.

He told me he had managed to escape his collar and was incredibly indignant about having it put on in the first place. He told me he wasn't just any cat; he was a wild animal, and he needed to feel free. He needed to roam, hunt, and there was an incredibly regal air to his energy. He was extremely offended by the collar — not just by the collar itself, but also by the sentiment behind it; the monitoring and the lack of trust as he saw it.

I told him his humans were worried and that they wanted him to go home. He said he knew they were looking for him, but he didn't want to go back because he didn't want to have the collar put back on. He said he would consider it if they promised not to put the collar back on. After thanking him for his candid honesty, I pondered on how I should pass on this information.

I spoke to the lady again and she confirmed the cottage I'd described was her house and her next-door neighbour had two golden retrievers. I just told her what had happened, believing honesty was the best policy, and I didn't want to change any of the message; after all, it wasn't my information, I was simply the messenger.

The lady was just fixated on how valuable he was and that he would have to wear the collar. When I later told the cat that they couldn't make that promise he wasn't impressed. I never heard if he went home but his message was clear – his quality of life shouldn't be measured in currency.

Louise and Barney

My good friend Louise wanted to check in with her gorgeous Black Labrador Barney, to see how he felt about a new human baby arriving in the family. The first thing this beautiful boy showed me was an image of him looking down on a family sleeping, all in one bed. He was looking over them for protection, but also as a sign he wasn't actually in the same room as them all. He didn't feel he could do his job properly from where he was sleeping. Louise said that was correct, he slept downstairs. From there Barney started showing me rabbits, I felt he loved them. Louise told me that he loves to eat them but it upsets his stomach, so he doesn't have them any-more! Barney showed me he loved walking around the trees, and there was a fireplace he's particularly fond of laying in front of. Again, he showed me him looking down on his family as a protector. The family is sleeping peacefully knowing he is there. Louise only had one child at the time of the communication, a boy. She is shown cuddling her first born. Dad is cuddling the baby girl, a new addition, but it was known she would come.

Dad is cuddling the youngest baby as he also takes protecting his family very seriously. He is in dark blue which represents order, strength, safety, and security. Louise is in pink, a deep pink which represents the strength of her maternal instincts and love for her family.

The blanket is purple, representing healing. It is laid over the whole family, healing past hurts. This family unit is a dream come true, and has helped heal old wounds.

The tree is for the love of the outdoors, and it has deep roots representing the family tree and the strength in its foundations. The bed is green, grounded, and solid. The wild flowers represent beauty, resilience, and longevity, and encompass this strong family unit.

Louise, Georgia, and Ava

Louise kept her horses in the same village and asked me for a communication visit with her beautiful Welsh Cob, Ava. I didn't know Louise at his point, so this was the first time I had met Ava and her other gorgeous horse, Georgia.

I didn't know any of their history but it soon all came tumbling out from Ava like a child trying to explain something through stilted sobs. Ava was holding onto a lot of emotional trauma and physical pain. She was also struggling with finding her place in Louise's heart, as Georgia was already fully installed!

There wasn't any jealousy, as animals do not feel this emotion, but there was a sense of sadness and loneliness at feeling slightly on the outside of the family unit. The other overwhelming concern for Ava was about Georgia; she was worried about her health. Georgia was in the next stable and although a little grey around the edges, her eyes still shone, and she seemed happy enough.

It also came to light that Ava has been labelled 'naughty' and 'dangerous' because of some extreme behaviours that had resulted in quite a nasty accident for Louise's daughter. I was now able to explain to Louise that these incidents had all been pain-related, and in that moment, I saw Louise's face change with emotion. Louise is a nurse so is used to dealing with such things, but this was so much more; the professional walls had come down and her emotions were now showing, and her heart opened in front of my eyes to also make room for Ava. By her own admission, Georgia was her heart horse, and Louise hadn't realised how much of her own emotional turmoil Ava had picked up on.

As I gently soothed Ava she came and stood right in front of me, literally with her face right in front of mine. She went into a trance-like state and much like dogs whimpering in their sleep, she kept making these little whickering noises with her eyes shut. I gently reassured her she was safe, she was loved, and that her physical issues would be addressed. She remained in this trance-like state for approximately half an hour and when she came around, she stretched and yawned for another eternity.

I hadn't realised it then, but this was the start of a wonderful friendship I now have with Louise, and I am blessed to now call her one of my dearest friends.

Not long afterwards Louise contacted me again to ask if I could go back, but this time to see Georgia. Louise was worried about her as she had taken a sudden turn for the worse and was struggling to get up and down. I knew without being told that this was one of those visits, and although I knew it was going to be incredibly emotional, I am always incredibly privileged to be able to help in these circumstances.

The visit was one filled with so much love. I sometimes feel like an awkward intruder when people ask me to ask that question, but Georgia was just so filled with love and thanks to Louise for her fabulous life, that she was quite sure she was now ready to go to sleep. In such sadness and loss, as we see it, there can also be such beauty in an animal being able to communicate this last request, and for a human to be able to give that last act of unconditional love.

I left Louise with Georgia and a couple of days later she messaged to say Georgia had gone home peacefully.

Louise's Words

I had met Ginny earlier in the year when I needed help with my mare Ava. So, when my elderly mare Georgia began to show increased struggles with daily life, I contacted Ginny.

Unfortunately, Georgia deteriorated and when I contacted Ginny again to let her know I needed the communication sooner rather than later, she came the same day.

Georgia told us she was ready to go to sleep. I was devastated, but knowing that I was keeping her here for selfish reasons lessened the guilt I felt.

Georgia went to sleep on a bright, sunny autumn afternoon, full of apples, and surrounded by those who loved her. I will be forever grateful to Ginny for allowing Georgia to tell me her wishes and doing what was best for her.

Mum's Words

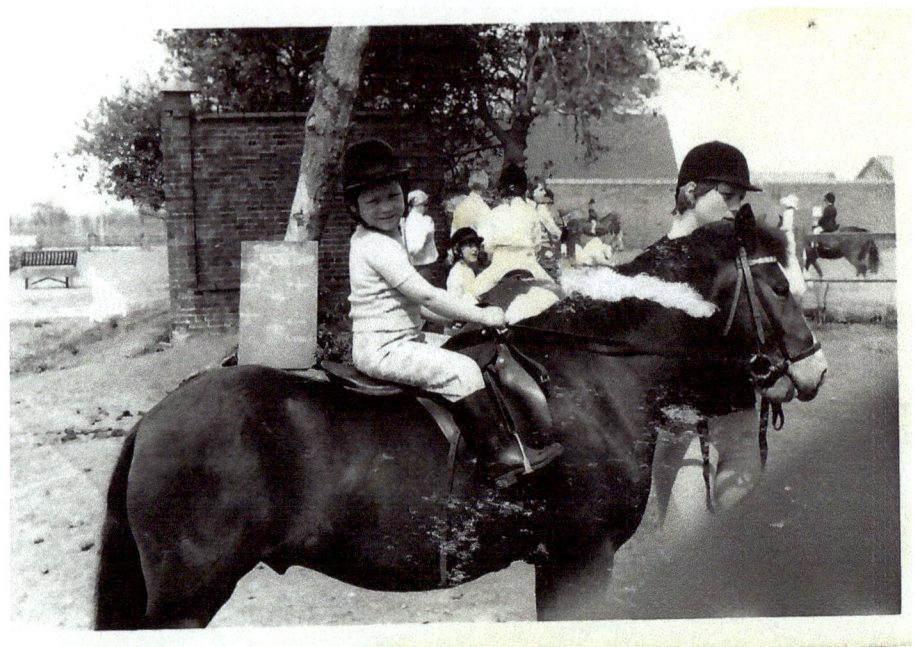

From an early age, Ginny had an affinity with animals. She had a little bay pony called Tommy Tucker, and enjoyed lead rein gymkhana classes held at the local riding school. She did quite well in the classes, mainly because Mum could run, Tommy was obliging and she could hang on tight, and we had a lot of fun! When Ginny was young, she always accompanied me to the stables, wrapped up in skiwear in the winter, and either sat in the car on the yard or played around with Humphrey the resident donkey. I then got my own little yard close to our house and bought a youngster. Ginny came with me whenever she was not at school. She loved being with the horses, even in the cold weather. As Ginny grew up, her love of animals did not wane. We had dogs, cats, horses, and rabbits, and even a couple of goats who Ginny loved to bottle-feed.

Moving on to adulthood, Ginny rode a friend's mare who she absolutely adored. Milly was a very special girl, and she and Ginny were completely in tune. I was later able to get Milly on loan and presented her to Ginny as a surprise, with a large bow around the stable one Christmas. Very sadly Milly got colic and needed to have surgery. She got through the procedure and recovered well, but a year or so later got colic again, and Ginny had no choice but to let her go over rainbow bridge. Ginny was absolutely devastated and rushed out to get another horse to try to fill the void.

Unfortunately, the horse she brought home was not a Milly. At that time Ginny was still grieving for Milly and was not strong enough mentally or emotionally to deal with Tizzy's more sensitive nature, so Ginny made the decision to exchange her for something more suitable.

My friend and I went with Ginny to see Skippy, after she'd been introduced to her with Mary the previous day. There were rows of stables with various types of horses hanging their heads over the stable doors. We peered in at them as we walked along, looking for this big, orange horse Ginny had been so excited for us to meet. Ginny went ahead and raced up to a chunky chestnut and just announced, 'I want this one.' My friend and I looked at each other and exchanged a quizzical look. However, a deposit was paid from money borrowed from the petty cash tin and Skippy, renamed Saving Grace, came home the next day.

Grace taught Ginny a lot and they had a wonderful time together over the years.

I am so proud that Ginny has discovered her gift of animal communication, and has been able to help so many horses all around the world by being their voice.

Aunty Mary's Words

Ginny has always been a little different to most children and young people where horses are concerned. I now realise it's a feeling she has for them that doesn't include winning, pushing, or ambition to squeeze every bit of talent out of them, but to work with them and appreciate how privileged we are to have them in our lives, and what they give us both physically and mentally. I remember the day we visited the dealers' yard with her confidence at a low, the attractive horses were all brought out and put through their paces, but she had absolutely no interest in trying any of them. In what seemed like desperation, the dealer shouted to her assistant, 'Go and bring out Skippy.'

From around the back, this chestnut mare was led out and although I didn't think she was blessed in the looks department, Ginny thought she was the prettiest, gentlest creature ever made, and a partnership was formed. Ginny hasn't had the best of luck with the health of her horses and has always been brave enough to make the right decisions, at the right time. She has recently found that she has a gift to communicate with these amazing beings, and is appreciated and praised wherever she goes, being called a saviour by her clients, and has helped people make what are sometimes difficult decisions with more ease and peace of mind.

Ginny now has Orion as her soulmate. He has carried on her teaching, and hopefully they will have many more years together enjoying each other's company

Thankyou Ginny for all you've done for Nic and Snip, and all equines and their keepers, and long may your journey continue.

Aunty Joy and Caily

Joy's Words

Caily (Eyla) was adopted from Many Tears Animal Rescue, Llanelli.

Her DoB was unknown but assumed by Many Tears to be a three-year-old, Pedigree Golden Retriever, actual Pedigree unknown. Eyla was removed from a breeder in Wales, along with other Goldies, by Many Tears on 15 August 2023.

Our little girl was adopted by us on 3 September 2023. Her name then was Eyla. On arrival at MTAR, she had an enlarged right eye, which was thought to have been caused by a 'trauma' at the breeders.

After consulting with the MTAR vets, it was decided that to give her the best chance the eye should be removed. We picked her up from MTAR on Sunday 3rd September 2023 (complete with antibiotics and the dreaded lampshade!) and considering she didn't know us and was still recovering from serious eye surgery, not to mention the fact that she had only ever lived in a kennel and run and had no experience of life outside that space, she travelled the five-hour journey home without a murmur. On arrival, she immediately made herself at home in the garden, but kept her distance from us and would not enter the house - we had to physically catch her, pick her up, and carry her to get her to come indoors for the night. This was stressful for all of us, but once she was in, she settled.

In a few days her eye operation scar was almost healed, and she was beginning to bond with us, even being brave enough to take a couple of steps into the house during the day, but we still had to manhandle her in at night. Another day or so and she was asking for cuddles and was coming into the house as far as the lounge, even during the evening, but was still refusing to come indoors after her late night wee.

I asked Ginny if she would communicate with Eyla to reassure her that she was safe but she needed to sleep indoors in future. Ginny communicated with her a day or so later about this and other things and, amazingly, that very night she finally decided to come in at bedtime voluntarily — and has not looked back since.

She is now called Caily, as she told Ginny she didn't like her old name. The young man who went through the adoption process with us at MTAR was called Cai, so it seemed right to use his name and just add the 'ly'. Caily is adorable, with a beautiful temperament, and is now well and truly one of the family. Her missing eye doesn't bother her at all.

Thank you, Ginny for helping us get through that very stressful time. It was only a little while after your communication, but her demeanour was completely different. She was obviously listening to you.

Hiding from the truth is something we have all done. Being open to communication means accepting there may be things that come up that may not make easy listening. The truth isn't always pretty, but what we want to hear and what we need to hear can be two different things.

I tell all my clients that I can't flower the messages up to make them sound prettier, because the true meaning and the intention can get lost. Sometimes the bare facts from the horse's perspective are exactly the pieces that have been missing from the puzzle.

This is where we must let go of our Ego, our Baggage, and our defensiveness, and instead of making excuses or trying to justify our actions, we must simply listen. Listen to our horse and their perspective. Then, we can plan to move forward so everyone benefits.

No matter how much I try to explain about communication, not everyone will completely understand it, and that's ok. Whilst some say they're completely onboard, their actions sometimes show that they only take the bits that suit their needs, and they are still able to ignore the bits that don't suit them, hence the blindfolds. Ego still plays a part even if it's with a little 'e'.

There is no judgement from me, just an acknowledgement that everyone is on their own journey with lessons to learn. The lesson for me here, as I sit with Orion and Spirit, is just to stay in my own lane and create my own boundaries for protection. Although they are trying, they still don't quite understand. Embracing a calling of animal communication is very different to expecting a quick fix.

Some livery yards I've visited have taken more self- protection effort than others. Here we see me working with a horse in communication. We are surrounded by pink and purple clouds for love and healing. The clouds get darker as liveries gather to watch, comment, and criticise my work. My human wants to have the communication to see how she can help her horse, but feels so much peer pressure and has suffered much negativity from the yard's huddle of self-proclaimed equestrian experts. My human feels intimidated and restricted by yard rules and politics, and feels like she has to defend how she looks after her own horse to this exclusive club. She feels that neither she or her horse have a voice, hence the muzzle and restraints.

When I'm working in communication, the humans who have asked me to come are also in my protection bubble but not in my communication circle. Blocking out all the negativity ensures communication can come through.

Energy can be affected by just one negative person taking out all their frustrations on everyone else.

But here we see how love and positive energy can combat and block those dark energies, and stop the life being sucked and drained from them.

When we feel the negativity affecting us, we can simply go to meditation, breathe, protect and ground ourselves once again.

Jodi and Dolly

Jodi's Words

I am the privileged carer of a beautiful seven-year-old mare. She is a traditional Irish cob, and a beautiful palomino with a sweet but sassy temperament. She came to me when she was five years old, and I was 65. As the second pony I have ever owned, she was a total re-education from my first, who was 16 when she came to me, with beautiful ground manners.

The first year Dolly and I shared proved to be a very testing one, both because I was really struggling to manage her in-hand, but also because she showed over time a recurring behaviour which I started to suspect was a health issue. Regular six-weekly physio sessions always found the same problem, and 'fixing' the problem only lasted 4-5 weeks before she started to evidence the problem again. During this period of about six months, I was introduced to Ginny by someone at the livery yard I was based at. From the time I found out that animal communication was a real 'thing', I have always been very open to it, so out of curiosity I contacted Ginny and asked if she could come to the yard and have a go at communicating with Dolly. At the point that Ginny arrived, I was starting to turn the corner on the difficult in-hand behaviour – I realised that I could not treat Dolly in the same way as I had my first horse, as a friend, but needed to show her some leadership to give her the confidence she needed. I had changed the way I worked with her and was already starting to see some benefits, although very much work-in-progress.

So, when Ginny arrived, I didn't (at her request) give her much info, as we were going to let Dolly feel free to speak on whatever she wanted.

The information Ginny gleaned from Dolly was extraordinary – and covered many aspects. It showed that Dolly's attachment to me was very strong but was causing a problem for her as she was taking on a lot of my emotional upset at the time (nothing to do with her at all), and it was causing her some physical stress. Ginny helped her release it, and at the same time helped me understand how I needed to be careful not to overload Dolly with my emotional issues. Ginny's conversation went onto how Dolly was coping in her environment in the yard – turnout and stable, and it confirmed for me where Dolly was struggling with her winter turnout, and, surprisingly, that she was aware of tension between myself and the lady whose horse was stabled next to her. Of most importance in that conversation, Ginny was able to let me know how much Dolly appreciated the changes I had made in handling her, by providing more leadership, and also that she wanted a vet to look at her because of the pain she was in.

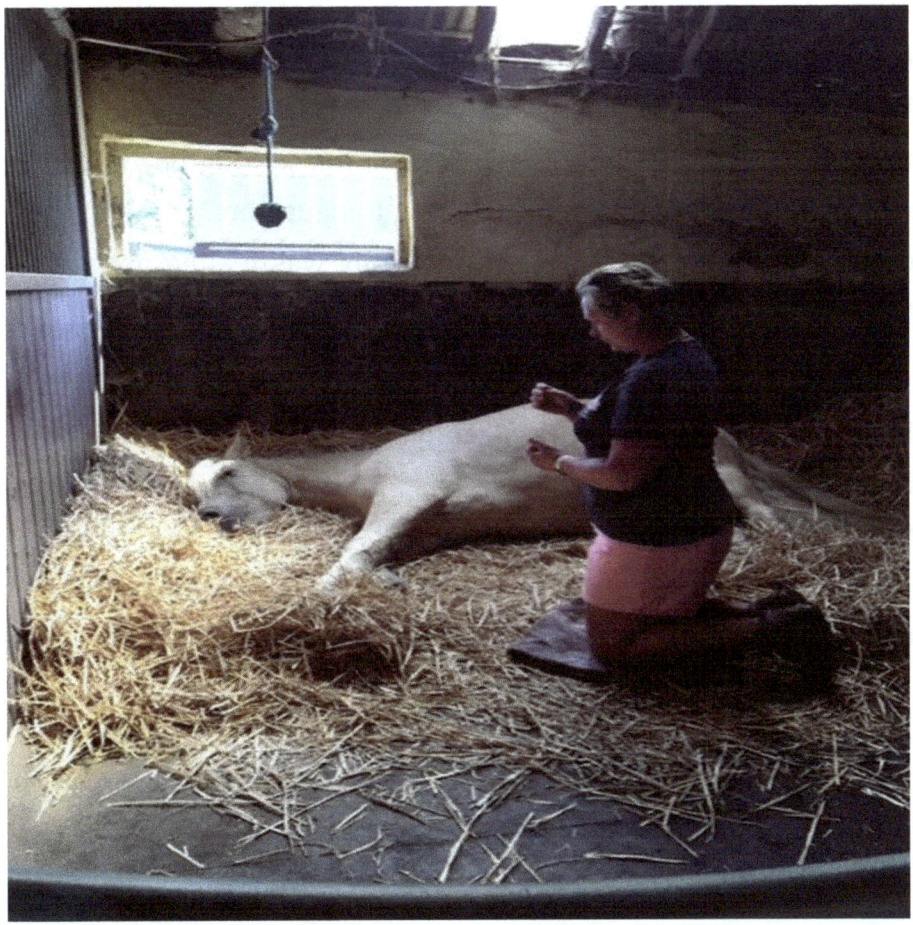

After vets' diagnoses that she was lame in her left hind (very subtle) and investigations ending in a steroid injection to both sacroiliac joints, the lameness disappeared, but x-rays had not identified what the actual problem was. Although the lameness had disappeared, I felt she was still not quite right, despite what all the 'experts' at the yard and the vet told me, so I asked Ginny to do a phone session, and she confirmed for me that Dolly was still in discomfort. Based on my growing confidence in my own instincts, and supported by what Ginny told me Dolly was saying to her, I pressed for Dolly to go to the Equine Hospital for a bone scan, and, thankfully, finally got a diagnosis of the actual cause of the underlying problem, which was arthritis in both sacroiliac joints, and that although giving her the steroid injections was the right thing to do, it had been going on for so long her muscles were spasming, therefore continuing to cause her problems. 20 days of muscle relaxant later and Dolly was feeling great and totally pain-free.

I know that if I had not had Ginny there to help me understand how Dolly felt, I would have really struggled to cope with the livery yard 'experts'' view that I was making a fuss over nothing; that Dolly was being a 'naughty five-year-old', and so on. I cannot imagine ever having an animal and not working with an animal communicator, preferably Ginny!

Seeing so many horses and environments with this work, it was hard not to notice that even though we are predators and horses are prey animals, herd and human group behaviour matches in regard to survival of the strongest.

Here we see a young girl sitting alone on a bench, watching while all the other children play together. She is a kind and gentle child but the others are ignoring her, playing together, leaving her feeling isolated.

Then we have a herd of horses. There is a hierarchy here and the little grey is being kept at the back of the line, away from the hay, at least until the others have had their fill.

There are horses who have told me about having back pain from an ill-fitting saddle. Not only is this giving them physical pain, but they are suffering emotional discomfort because of the physical pain.

As humans, we must put aside our Ego and even if we have a competition booked, a training event, or a pleasure ride booked and paid for, we must consider the horse first. Here we clearly see red under the saddle for pain in that area. Once the saddle is off, we can see there is also stomach pain and a headache. The back pain is causing other physical and emotional pain, which has been noticed but ignored here by this rider. The rider is pictured in a yellow cloud, they are having a lovely time and haven't wasted their entry fee – but at what cost? Human Ego is often responsible for some hefty vet bills.

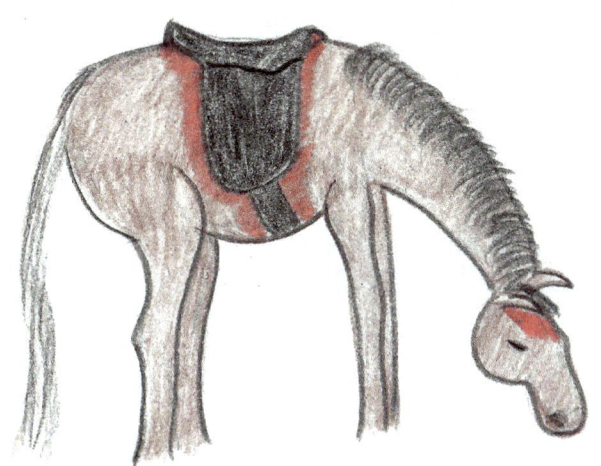

Would we send our children to school in shoes that didn't fit?

One of the most beautiful things with Animal Communication is being the voice for the horse and passing their information to their eagerly awaiting human.

Here, I am dressed in pink for love and purple for healing; I have communicated with this beautiful horse surrounded by a purple cloud for healing through Spriit. I am passing on this information in the form of Spirit to the human who is dressed in orange for vibrancy.

The human then, in turn, takes that information, surrounded by a pink cloud of love, and uses it for the good of their horse, strengthening their bond and relationship.

However, there is also a lot of frustration when I pass on the horse's information and it's not what the human wants or expected to hear. The information doesn't fit in with their plans, so essentially nothing changes. Now, I know that not all requests can be catered for, but this is about the ones that can be but aren't, because it doesn't suit the human.

Here we see me in communication and passing the information on. The human doesn't like what they hear, so I am left feeling helpless to assist the horse any further and apologising for my species, yet again

Natasha and Zeb

This was a first for me: a photo communication where the horse made it clear, halfway through, that he wanted me there in-person to continue the conversation!

The handsome Zeb had some emotional trauma from a previous home that he wanted to offload, and helped me explain to Tash some feelings he had about his environment. A very special relationship where, again, horse and human are healing and guiding each other.

Natasha's Words

I have had Ginny come out twice to see my boy Zeb. I have owned him for four years but felt there was something there that I was missing with him, after several months of deliberating if indeed having a 'communicator' come out would help? On the same night, Ginny popped up on social media and I knew it was meant to be! It was the best thing ever; she enabled Zeb to offload an enormous amount of emotion that he was carrying around with him, made me understand him more, and our relationship has grown so much more. Can't thank Ginny enough for her time with Zeb, and made me believe 'I am enough'. Thank you, Ginny, for who you are!

Whispers in the Wind

After my training, Richard kept in close contact to help and mentor me with any questions or situations that arose with any of the animals or, indeed, the humans!

I was deep in meditation one day when this image came to me, so I thought I'd attempt to draw it for him; my way of showing Richard how I saw his new song title *Whispers In the Wind*.

A story of his life and of his love for his horse Willow, and their shared journey together. I was honoured and amazed to have been asked and felt totally safe in Richard's trust, and his confidence in me and my new-found ability.

When I listened to the song, I was totally engulfed in the story and the emotions it evoked within me. Richard's emotions became my emotions, and I was transported to so many different places Spiritually and emotionally that the sketches just poured from me. I just couldn't stop.

Drawing after drawing appeared on the pages in front of me, the messages so strong and images so poignant and symbolic. I have never felt such a strong emotional connection to someone in this way before. It was almost like I could get inside Richard's head and heart just from his lyrics.

Of course, we talked for hours and hours about the drawings, and there were so many that Richard said he would like to use them to create the video for his song, as well as the CD cover!

Through this shared venture, we became very close and our relationship turned into something much deeper. This collaboration was something very intimate and became an incredibly special project. It was because Richard had been so honest and vulnerable with his story that a strong and beautiful friendship flourished, with me helping bring his song to life.

Richard then asked me to illustrate his next song Little Bird, which is all about the robin who had featured in Whispers in the Wind. It tells the robin's story and his fight through adversity.

I think we both helped heal each other, too. Richard helped me with my confidence by believing in me completely, right from the start. I hadn't realised this was something that had been missing in my life; true faith in me, no buts or how I could've done it better. I was enough as I was.

Finally.

Unconditional means absolute, and not subject to any special terms or conditions.

Animals all come from unconditional love. They don't have any agendas, or rules for how they love, they just do it.

Human relationships are rarely unconditional. Even love for our own family members isn't necessarily unconditional, but I do know some people who say that they do love certain family members, even if they don't always like them!

Even marriage comes with conditions and rules we must follow, to stay within the parameters of what society has deemed as acceptable boundaries for this alliance.

I know most people need to justify the expense of keeping horses by how useful they are, and I understand that from a human perspective.

What is difficult, is to explain to some humans that their horse knows when their relationship depends on certain things and is not unconditional.

I will love you if I can ride you. I will love you if you are healthy. I will love you as long as you behave.

I will love you as long as you serve me in some way.

It's very difficult to have a trusting relationship when these conditions depend on how you will be cared for and where you will live. I have seen so many heartbroken horses who know their place in the family unit has such expectations, and that sometimes they just can't live up to them.

The size difference isn't just physical.

Who is saving whom? I think we can all relate to this in some way.

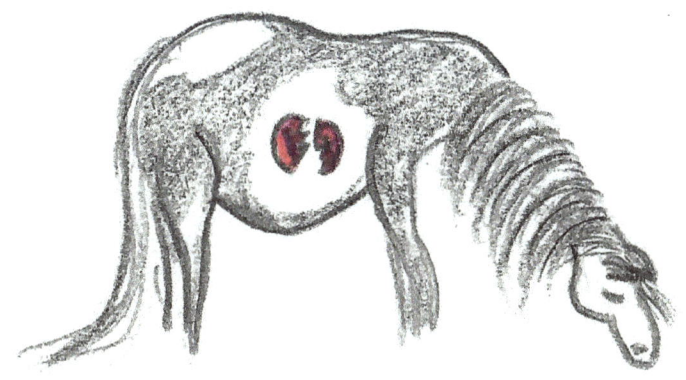

So much trauma and separation anxiety in horses comes from their weaning process, both in the mare and the foal.

Every mare I have communicated with, who has had at least one foal, still carries emotional trauma from being separated from their baby before they were ready.

Depending on the method used, the trauma scale varies from outright grief to sad acceptance about their missing offspring.

The stud mares whose babies are taken quite early but are weaned in groups, do feel like baby-making machines and do not like being forced to have babies every year. Having babies isn't the issue for them, it's the fact they would breed every two years in the wild, allowing them to wean the foals naturally themselves.

The home brood mares are the ones that appear to suffer the most emotionally. Again, the foals are taken away far too young, and then they are expected to trust the human who has taken their baby away.

The novice breeder, those who have never bred a horse before and don't understand all the risks and potential pitfalls. The mare has done well in competition, so they think they are giving her a reward by letting her have a foal. How are they going to explain to the mare that she can only be this lovely mum for six months, then they plan on selling her baby for a good price.

The happiest horses I have seen have been the ones who have been bred and stayed in the same home as their horse mum.

The mare is happy and willing to go back to ridden work and still has a strong relationship with her human, unlike those weaned by being locked in the stable with the top door shut.

The foal is happy and has no reason to distrust their human.

One of the most loving and memorable communications I have experienced was with a beautiful Thoroughbred brood mare who was blind.

She had 'exceptional' bloodlines (they all do in my eyes but that's a different conversation!), so her physical difference meant her job to benefit humans took a different turn than what was originally expected. This gorgeous girl loved being a mum and apart from doing it every year, and having her babies taken away too soon, was enjoying being a horse as much as one can in domestic servitude. Each of her foals had a bell around their neck so she always knew where they were.

I have mixed feelings about these situations, especially when the mares are expected to breed well into their teens and twenties, but overall, a happy horse and a mum in a million.

There is no such thing as a 'naughty horse'.

A horse isn't naughty. A horse is just behaving like a horse.

A horse is just behaving like a horse, but in that moment, they may not be demonstrating the behaviour the human wants. That doesn't make the horse naughty.

The horse is just behaving like a horse.

The human likes to justify their own actions, especially when they are harsh or involve inflicting pain or dominance, by putting human emotions on to animals.

Anthropomorphism does not make naughty horses. Horses reacting to a negative experience does not make a naughty horse. Horses engaging with us does not make a naughty horse.

Thousands of years of DNA cannot be completely eradicated from evolution, just because the human wants the horse to behave in a certain way to suit their own wants and needs.

The 'show them who's boss' brigade show only their ignorance, as horses are generally reacting adversely due to fear, confusion, or pain.

Partnerships and relationships work with good communication, not from a dictatorship.

I'm very cautious about using the term 'Natural Horsemanship'. In reality, whilst some of the behavioural issues can be dealt with using a mixture of these techniques quite successfully, they don't always address the emotional issues behind the behaviours, or why they are occurring in the first place.

Let's be honest, there's nothing natural at all about riding horses; they aren't designed to be ridden but are conditioned into accepting it.

Generally, horses in domesticity cannot make any choices for themselves. Everything about their life and welfare is decided by a human, even if the horse knows it's not what they want or is what's best for them.

In the Natural Horsemanship world I experienced, there seemed to be an obsession with teaching horses to lie down. I know for some it's a bit of a party trick, and whilst every care is taken that the horses aren't hurt physically, symbolised here by the use of knee pads, I cannot say the same for the emotional side. The whole point, it seems, is that in forcing horses to lie down and submit to us as a dominant species, it apparently makes them feel that they have no choice but to follow us and see us as their leader.

I can see how it may wow the crowds on a physical level, but personally on an emotional level, it breaks my heart.

I don't feel it should be seen as some magic trick to show off to your friends, as that's just coming from a place of Ego. For me personally, it feels degrading, violating, soul- and heartbreaking, and that's because that's what has been shown to me when I've seen it happening. One time I was left sobbing uncontrollably from the sheer feeling of humiliation and helplessness.

There is a distinct difference between dealing with behavioural issues and communications, and it's important to understand the difference.

All horsemanship techniques, classed as natural or otherwise, have all come from *a human's perspective of a horse's perspective.*

Scientific studies, behavioural observations, even wild horse studies, are all written from a human's view and understanding of what they have witnessed, studied, and concluded.

What we do as communicators is relay the information directly from the horse; from their view, from their experiences, and from treating them as individuals, which of course they all are.

Often, after a communication, the horses behaviour changes through conversation and understanding, without the need for pressure halters, sticks, or flags.

For me personally, any kind of pressure, be it a double bridle or a halter, is still force unless the horse has a choice to say, 'No, thank you.'

And when that happens, how we respond to that, will depend on our Ego and Baggage.

Of course there are times when the horse must do as we ask; notice I say ask, not instruct. But, if we treat our horses consistently with respect and give them choices, and more importantly respect those choices, they do the things we ask of them much more willingly.

If our horses turn away from us when we approach them in the paddock, how many of us stop and wonder why that has happened? How many of us just carry on and catch them anyway?

How many of us actually check our energy before we even approach our horses?

There are so many communication subtleties that get lost in our hurry, in our need to get the job done, to ride quickly before the weather changes, to think about what is for supper — all times when we aren't with them mentally in the moment.

How do we like it when we are trying to talk to someone and all they are doing is texting or scrolling on their phone?

Horses live in the moment, so when we are with them physically, we really need to remember to also stay with them emotionally. We know how it feels to talk to someone who is constantly distracted and isn't really paying attention to us, and long term, it can be really damaging to the relationship.

Because communication work is very draining emotionally and physically, I tend to limit visits to my local area. But as communication works with a Spiritual connection, I am still able to communicate with animals when I'm not actually with them, either by photo or via live video link. This has enabled me to help horses all over the world and so far, I have clients in Belgium, Holland, Australia, and Gran Canarias, as well as all over the UK. This also applies to mediumship work — connecting with animals passed over — but currently, I don't feel drawn to this work. My strongest connection is with those animals still here in the physical, but as my skills develop, it may well lead me there. I wait for Spirit to guide me where I'm needed.

Helen and Aquarius

Helen and I have been friends since we were teenagers. She was Head Groom at the yard where Milly had lived and has also given me lessons over the years. We were even neighbours at one point, but now Helen lives in Belgium with her partner Johan. Helen is not only a talented showjumper but has always been emotionally invested in all her horses. Not long after my courses with Richard, I asked Helen if I could try doing a long-distance communication from a photo with one of her horses. I didn't need proof or evidence that this stuff really worked, it was more for my confidence, and as Helen didn't live in the UK, I felt it would be good experience, without too much pressure, to see how my skills were progressing. Always up for anything to help her horses, Helen readily agreed and sent me a photo of one of her horses, Aquarius. Helen confirmed everything that Aquarius had said was spot on and identified the gold star that she had shown me mentally as her other horse Goldfee, their 'top, star' showjumping horse! Aquarius talked about wanting a baby in the future and showed me an image of a horse she really liked who wore a bright blue rug.

Helen couldn't contain her disbelief and amusement when she heard this and sent me a photo of the new stallion on the yard. Aquarius had identified her baby daddy!

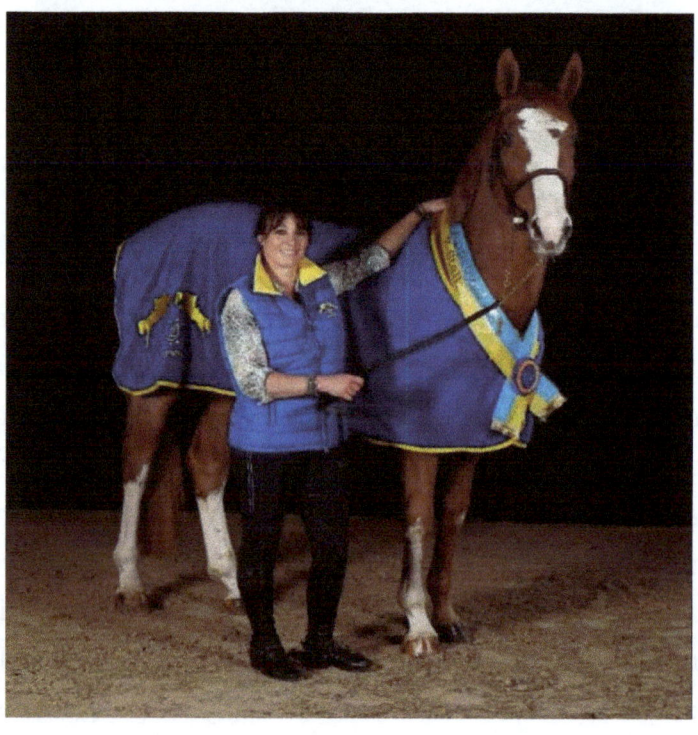

Helen was so pleased with how things had improved for her and Aquarius since I communicated with her, I was then trusted with the yard's top superstar Goldfee!

We did a live video link with Goldfee, who had some personal requests regarding a feeding preference, and she also shared some very emotional experiences. While I was doing my meditation warmup prior to calling Helen, I had a very strong image of Joey from War Horse, the theatre production. I wrote it down and mentioned it to Helen later. Towards the end of the communication, I asked Goldfee about it. She then went on to tell me about how she had been in World War I in a previous life and how it had been a very difficult and hard time, and she carried that work ethic with her into this physical life. Helen told me that where they live is right where those awful WWI battles took place, which I didn't know!

Helen's Words

I've known Ginny for over 30 years both personally and professionally. Where horses are concerned, Ginny has always put their happiness and welfare first, and has really found her calling through communication.

I am based in Belgium, so it was a perfect opportunity for Ginny to use one of my horses as a case study for long distance communication. Having no previous knowledge of the horse, Ginny quickly formed a bond with her and was accurately describing situations, feelings, and images relating to her competitive and everyday life. After becoming aware of some issues and making the necessary management changes, the difference really started to show, and our relationship strengthened.

We have since asked Ginny to communicate with other horses and have no hesitation in recommending her as a communicator and voice for the horses.

Helen xx

After my first full year of working as an animal communicator I learnt and grew so much with each new client I met. I'm so grateful to all of them and just wanted to share a few things I discovered about myself and others during that time.

- There is a fine line between knowing your own worth and letting Ego slip in

- There is a fine line between educating and preaching

- Who makes me all-knowing anyway? Well, the animals do. It's not my information, it's theirs, and they are all-knowing about themselves

- Who says I'm truthful and others could do things differently? Well, again, the animals!

- How far down the education route do you go when it falls on deaf ears?

- There is a fine line between helping and interfering

- There is a fine line between doing nothing as a bystander and staying in your own lane

- There is a fine line between putting yourself out there but not pushing yourself onto anyone

- There is a fine line between waiting to be asked and appearing like you're not interested

- There is a fine line between the truth and offending people

- There is a fine line between self-reflection and seeing the truth

- There is a fine line between seeing the truth and being defensive

- There is a fine line between Ego and arrogance, and knowledge

- There is a big line between doing things with your horse and for your horse

- There is a bigger divide between human emotions and behaviour than animals'

There are many fine lines to navigate in life but if our true intentions are for the good of our animals, away from human wants, Ego, pride, and defensiveness, then we'll get there.

Yes, the truth can be difficult to hear but, equally, what better foundation to renew our relationship with than truth and love? The truth is also a beautiful thing to hear, especially when our horse trusts us and wants to tell us how much they love us!

What better way to show our horses that we care than by listening! Even if there's a grumbling suspicion about something, we can plan to move forward and make life so much better for both, rather than just turning a blind eye and hoping things will improve while ignoring the subtle signs.

It's been very humbling and emotional but so, so fulfilling.

I'm often asked what I say to the sceptics. Those who pooh-pooh the woo-woo without ever experiencing the juju.

Well, I say nothing, nothing at all. It really is just white noise.

It's a little like when people say they don't like a particular food but have never actually tried it! Everyone is entitled to their own opinion and belief system. Everyone has their own life experiences, own conditioning, and Ego to deal with.

It's just a case of respecting each other and eventually, if people are interested, they will ask. If not, it's their journey and it's in no way a problem or issue to me.

Like attracts like, and Spirit will always find a way to guide those who are ready.

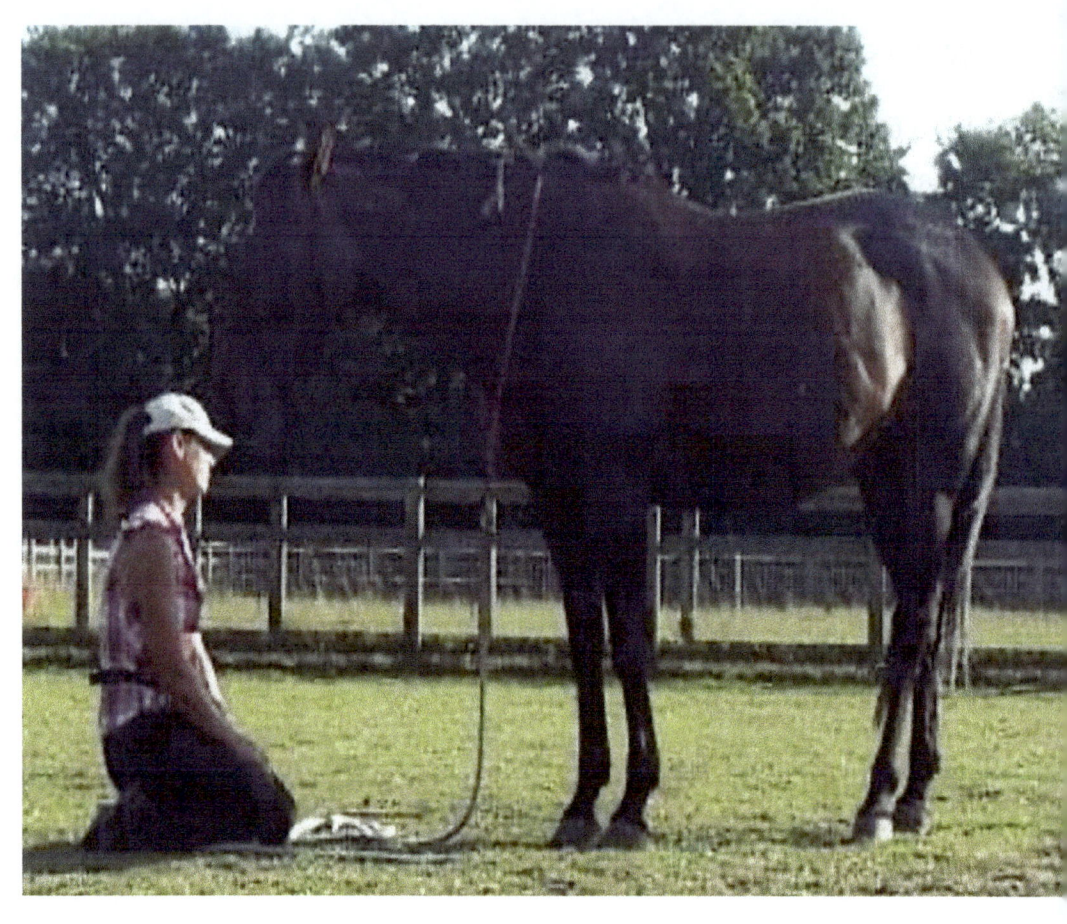

When the student is ready, the teacher will appear.

I understand that some people are not necessarily sceptical of my work, but are more scared that the outcome may mean they need to make changes. Those changes may not be financially profitable, and could mean that a complete change in their current horse management is required, which can be a big Ego dent to some.

One yard visit resulted in several more liveries wanting me to communicate with their horses. After a few days, word got back to me that although they were keen, the yard owner had expressed their opposition to my work and it was made clear to the liveries that I wasn't welcome back. Not wanting to upset the yard owner, the liveries decided not to go ahead with the communications. I was also warned to be prepared for some hostility, verbal challenges, and insults if I did return.

I have not returned to the yard but I would if I was asked. Everyone is on their own journey.

I don't take this personally, as I know this response comes from a place of fear; fear of the truth, fear of change, and fear of challenging a lifetime knowledge of horses. This is where Ego comes in, and if we truly want the best for our horses, we overcome all those fears. It also leaves me wondering how much influence a yard owner should have over the practitioners their liveries use, and if they should dictate that care package based on ignorance and fear. Of course, anyone abusing the animals or their humans should not be welcome, but to be so fearful of what had been a truly remarkable communication left me once more acutely aware of the vast gap between Equestrians and Horsemen.

This sketch symbolises a yard owner remonstrating about me helping a liveried horse when in fact their own horses are the ones in physical and emotional pain

Sam and Zimba

Sam's Words

I had the absolute pleasure of having the incredibly gifted Ginny visit my yard. What I experienced throughout the day is something that will stay with me for life.

My old mare Zimba went first, and she willingly accepted the 'connection' when Ginny asked. Without any doubt whatsoever the two of them were chatting. My old girl hasn't always been the easiest and I've spent a long time trying to figure her out. Through Ginny I learned of some emotions that she'd been carrying with her for a very long time, which she was able to let go of. I had always known that Zimba and I had a bond, but until then, I didn't know just how strong that bond was. We genuinely have an unbreakable bond. Zimba has often shown possessive behaviour towards me, and I learned that she in fact is showing protective behaviour towards me. I am her human. I have known Zimba a long time and she once was a slightly quirky and anxious little sass pot. It's nice to know that she feels safe with me and has been able to let her quirks and anxiety go as we've figured each other out over the years.

Ginny then spent the rest of her day working through the other horses on my yard. I listened in on each session and I know everyone got something incredibly personal and positive out of their experiences.

I believe 100% that every horse was communicating through Ginny. To be able to witness four different horses and the ways that they all reacted during the session was amazing.

For anyone who is wanting to book a session but is scared to, there's really no need to be scared. It is emotional and you will need tissues, but we all laughed just as much as we cried. I can confidently say that it was one of the most magical and love-filled days that my yard has ever experienced.

Thank you, Ginny, for an amazing day, and for being such a great interpreter for every horse on my yard.

Christine and Sidney

Christine's Words

Sadly, Sidney is no longer with us. He wasn't blessed with good health, and in December 2024 I had to say the saddest of goodbyes to him. People talk about having a heart horse; I believe he really was mine. We were brought together in 2015 and spent a lovely almost 10 years together. He was the gentlest soul but quite insecure and nervous, and certainly in the earlier days, rather prone to over-reacting. He taught me so much in our time together about patience, kindness, and encouragement, as well as giving me a fairly comprehensive walk through the entire veterinary manual! I showed him love.

It was several years into our partnership that we first met Ginny. The experiences that we shared really helped me to trust in the bond that Sid and I had. Our communications helped us both on so many levels; from the big things, like 'Does Sidney love me?' and 'Is he happy to be ridden?', down to the more specific, like 'Could I have a fleecy cover for my girth please?'. Of course he could! Fleece cover was provided before our next ride.

There are so many examples of how Ginny was able to show me ways of making Sid's life better. One example that springs to mind also meant that Sid altered his usual behaviour too, which really blew my mind! During a communication, Sid sent Ginny a feeling of grittiness in her teeth. On discussing it with Sid, it transpired that he felt a roughness in his top right teeth, but he didn't want a vet or dentist to see him.

He actually said, 'You can help me.' (meaning me!) How on earth was I supposed to help him? Ginny asked Sid if he'd like his teeth flushed out with warm, salty water, to which he said, 'Yes.' 'Mm. Ok,' I thought. This is a horse who has issues with his mouth. I had tried to improve the situation in the past with syringes of jam to try and encourage him to accept a syringe, but it hadn't really helped that much! When there was a need for me to worm him, I used to go into battle with my riding hat on and had learnt to untie him before hanging on for grim death as he whirled me around the yard, with me squirting wormer generally everywhere other than in his mouth. So, I think quite understandably, I was dubious. I told Ginny my concerns, and she said, 'Let's ask him.' She asked Sid, 'If your Mum washes your mouth out with the warm salt water, will you be brave and stand nicely for her?', to which he replied, 'Yes.' So, later that day, after we had finished our chat, I prepared the salt water. I brought the syringe up to his mouth.

And he pulled away. I held his headcollar and quietly said, 'It's ok, this is the warm salt water you asked for,' and gently put the tip of the syringe to the corner of his mouth. To my surprise, he softened and relaxed. I squirted a little bit just inside his mouth, then moved to put the whole syringe in, directed to the top teeth at the back and flushed. Then another syringe full, flushed. And again, and again. I must have rinsed his teeth with about eight syringes of water, and he did not move at all! I am quite sure that, without talking to him with Ginny, there is no way that would have been possible. To be honest, without Ginny, I wouldn't have even known that teeth flushing was a service Sidney was requiring!

Sid had a great sense of humour. He requested a change of bit, from the straight bar pelham he had been wearing for many years to one with a joint, not a single joint, but a roller type with a bit of flex. I asked him if he wanted to keep the curb chain, and he said no. He was a strong lad; a fairly chunky, 16.2hh Irish Sports Horse with quite a big canter, and when he got rolling there wasn't a great deal I could do to affect the speed. Stopping required a reasonable amount of preplanning! So, no curb chain was a concern. I asked him outright, 'If we get rid of the curb chain, do you promise to stop when I ask you to?' He said, 'No.' That did make me laugh! At least he was honest!

We explained to him that for safety reasons, it's important for me to be able to stop when needed, as it could be dangerous to canter into a road, or the horses we are with might need us all to pull up, so we agreed the curb chain could remain. Sometimes you can't have everything exactly as you want, but being able to explain why not I believe really helped him to understand life.

One of our favourite things to do was to go to Thetford for our little holiday once a year. After being ill in early 2024, Sid picked up well and had a lovely summer. We had a chat with Ginny in the late summer and before we ended, Sid sent her an image of a Mariachi band. Ginny told me and asked if there was a long-haul holiday looming, as this is one of the regular images she receives for this. I said there wasn't as we don't go abroad. In fact, I wasn't going away at all (other than Thetford in the next couple of weeks) and didn't know anyone who was. Sid persisted and kept up the Mariachi band image. Ginny and I were confused, but then I had a thought. Normally we go to Thetford for three nights, but this time we were going for four nights, as a treat. I told Ginny and she said that it probably wouldn't be that as it wasn't that significant, but she asked Sid anyway. A big 'YES!' To him a four-night holiday was a big holiday, which for him was definitely Mariachi band-worthy! He was such a sweetheart.

I really do miss Sid very much. I take a great deal of comfort from the memories of the lovely times we had, and having had our conversations with Ginny I feel like I was privileged to know him at a deeper level than many people get to enjoy. He was kind, honest, and funny. He cared about me and even had opinions and advice for me on friendships and family. He was incredibly perceptive and wise. I truly believe we were brought together for one purpose each; for him to teach me, and for me to show him love. Mission beautifully accomplished. Forever my gorgeous Sidney.

Christine and Chester

Christine's Words

'I've never been loved by a human.'

This is the key message that has really stuck with me from our first chat with Ginny. Knowing that Sidney was starting to struggle slightly physically, I thought it might be nice to get another horse. With Ginny, I asked Sid if he minded and he didn't, so I began looking around. After a while, this lovely looking chestnut ex-racer came up on an advert which had been forwarded to me by a friend. I went to meet him; he was nice, and a week or so later he came home.

He wasn't quite what I was expecting. The horse I met was relaxed and easy going. The horse that arrived wasn't. I was taken by surprise, and for quite a few weeks I really questioned whether I had made a mistake. I felt that he was a lovely horse, but maybe just too much for me. I started making all the usual appointments you do for a new horse (saddle fitting, dentist, sheath cleaning, physio, etc) and of course, booked Ginny. The session we had completely changed how I felt about this new whirlwind who had come into my life. He had a lot of physical pain and weakness, all of which were picked up by our physio a few days after our communication, without telling her any of what he had said! He was confused and didn't know he was with me for the long term. He had emotional trauma, having heard and seen distressing things. He had been treated roughly by more than one person in the past. He felt that he didn't need a human at all. I realised at this point I had a confused, lonely, lost soul standing in front of me. Stability, consistency, and love is what he needed, and I was the person chosen to give it.

Understanding this had a huge impact on me. Without Ginny being there to be a window into his world of feelings, I cannot be sure where we would be now. He is a very sweet boy, and we are now a few months down the line. Things aren't perfect, but life never is! He is still quite fiery at times, but knowing how he feels and what he has been through motivates me to take each day as it comes, and to continue to be his constant. The one who is there to feed him and talk to him every day. The one who holds his head when he wants to stand and bury his face in my coat to doze. The one who squeals when he nips my fingers, as everything goes in his mouth when he's feeling anxious. The one who will not give up on him.

As well as the significant revelations, after our session I also had a list of his requests, all of which I fulfilled – what would be the point of him saying anything if I was then to ignore it? A change of bit and a tummy supplement. A new name! He didn't like the name he came with and asked to be called Chester. He told us he found the way food was delivered to him was rude in the past (feed bowls chucked at him), so every feed I have made sure I place it down nicely in front of him, and the way he receives it is much more polite now too – he used to pin his ears back at every feeding, but now he has a happy face the vast majority of the time. Just small changes which can have a big impact; letting him know that I respect his feelings, and he is reflecting that in his actions too. There really is so much value in being able to have a two-way conversation.

Following the session with Ginny, reflecting on everything Chester had said, I realised he had already taught me two important lessons; 1. Don't place expectations on horses and 2. You might not get what you want, but it might well be what you need!

I'm looking forward to seeing what the future brings for us both, and having our next chat with Ginny!

Managing expectations

Communications aren't necessarily a quick fix for issues we may be experiencing with our human-animal relationship. It's more like a 'triage' process with the physical issues, and the animal will guide the human, through us as communicators, to get the correct professional to help them. We cannot and do not diagnose. Emotionally, however, that's where the magic happens, and we are the correct practitioner and that can be enough; however, the two are often interlinked. Whilst the physical care of animals is well documented, the mental and emotional health of our animals is so often overlooked or not understood. Look at how mental health is only now being considered for our own species! What we do as communicators is very different to behaviourists and trainers, as we work on a Spiritual level, but when we can all work together, that's when the animals get the complete care package!

This work is truly wonderful, but it is also incredibly draining both emotionally and physically. Communicators are humans, not machines, and we also have our own lives running parallel to communication work.

Sometimes we just have a bad day, are affected by the lunar cycle, or the energy isn't quite right to be able to give clients 100%.

Communication sessions aren't just the time spent with the animal, there is a lot of preparation ahead of that visit or long-distance photo or video call: planning, meditation, ensuring we won't be disturbed, creating the right energy and atmosphere before we even start.

To give you the very best of me I must look after myself first, so if the energy isn't quite right, I will rearrange the appointment.

Communications aren't like other treatments you may book, so it is very important to respect this, even if it is disappointing.

Richard rescued a beautiful mare named Ebony. An ex-racehorse destined for the meat market. Ebony is now thriving in her loving new home but unfortunately, not all ex-racehorses have such a happy ending. So many of them are discarded, broken physically and emotionally, having been used as money-making toys for our species.

Hold on sweetheart, he's coming!

Within the equestrian world there is an unspoken hierarchy. The competitive circle tends to look down on the pleasure riders and indeed the non-ridden community.

The irony of this is that the horses being ridden on roads and on our precious few bridleways, are far more worldly and have to deal with many more hazards and challenging situations than many competition horses would ever come across, or indeed be able to cope with.

We all know how expensive it is to keep horses and look after them well. I know I've gone without and still do, so Orion can have his essentials and have an emergency pot, which of course is never enough. When Milly had her colic surgery, I had to sell my car to cover the costs and, like many horse lovers all around the world, we do it from love, dig deep, and find a way.

I know many people have sharers to help with costs and the exercise regime and this arrangement often benefits everyone. But how many humans consider if the horse is happy with the arrangement? How many horses have multiple riders and how many horses are ridden in awful weather just because it's the sharers day and as they've paid, they're going to get their money's worth? How many horses generally have more than one rider?

It breaks my heart when I see horses being ridden out in weather in which I wonder if the rugs should even be removed, just so the sharer gets to ride. What does it really achieve? Is it enjoyable for anyone in such extreme conditions?

I used to do it myself, back in the days when it was drummed into us; I must ride my horse six days a week, 20 minutes is better than nothing, I must school in the driving rain and sleet as the horse must be ridden to keep fit. But not anymore.

It's not only the riding conditions, but also the constant adjustment to multiple riders. Their physical differences, their differing abilities, their differing emotions and energy, their differing expectations, tolerance, and acceptance levels.

I know the horses don't all enjoy it and yet again, it's another choice taken away from them, not only if they want to be ridden in those conditions but also by all the different people.

This is when I see so many horses shut down, and doing what they've been conditioned into accepting as their life.

Out of the hundreds of horses I have seen now, of those who had sharers, and of their humas who have thought to ask the horse if they are happy with the arrangement, only one pony said he was happy with the others who rode him, but even he had one rider he didn't want to continue with!

Let's look after our horses and be sensible about riding conditions at certain times of the year. Conditioning can distort our Ego and our assumptions about our horses, so if we ask ourselves why we really need to ride in 60mph gusts, the answer is usually 'human conditioning'-based. Rather than letting our Ego take on the elements, let's put our horses first, as we all know how that usually ends!

Vicky and Ciro

Vicky Lewis and her beautiful Ciro in Gran Canarias share a very strong bond and an incredible relationship that other humans sometimes struggle to understand.

Vicky has been, and still is, Ciro's lifeline, and after a tough start Ciro has struggled to make himself understood and his voice heard.

Luckily for him Vicky's instincts have served him well and saved them both as they continue their road of destiny together.

'Sometimes you don't get the horse you want but the horse you need.'

Beautiful things can happen when the details are addressed.

Vicky's Words

Such an amazing achievement for me to be back riding Ciro, after thinking I would never ride him again. Also, I'm leading him out in-hand with no striking out at me with his front legs, although he still does occasionally get me with his teeth! Not only did you help Ciro, but you helped me regain so much confidence that I had lost, by helping rebuild my relationship with him.

As communicators, we are not anti-riding or anti-competing.

Horses have had to adapt to living in domesticity and so rely on us for everything.

They cannot make any meaningful decisions for themselves so it's up to us to ensure they are looked after physically and emotionally.

Communicators help the horses' voices be heard, so that their truth can be heard, not from a science or human perspective, but truly from their hearts.

As communicators, we want horses and their humans to be happy, just NOT at the horses expense or at any cost.

A good relationship involves two-way communication, no matter the species, and a good partnership involves listening to each other. It should not be a dictatorship. When both parties are happy, truly happy, not just when human's expectations are being met, then everyone wins.

Retiring a horse should not mean forgetting about them
and not giving them any more care or attention. Non-ridden
horses still need all their physical and emotional needs to be
met, not to be just left in a field.

Oh good, its Saturday again!

We must think about who the birthday celebrations are
really for.

Juan and I got together in 2013 and married in 2017. Sometimes the biggest surprises are the biggest blessings! We both share the same passion for horses, which only comes with it being in your blood and when they are a lifestyle, not just a hobby. Its thanks to Juan and his passion and his experiences, that I have accumulated so much knowledge about polo, racing, breeding, and of course the horsemanship and Gauchos of Argentina.

I did this sketch, as without Juan I wouldn't be where I am now. He has learnt from me as much as I have from him, so together we make a great team!

Juan spent four years studying to be a vet in Argentina, but life had other plans for him. Before he could finish his final two years, he found himself in England as a polo groom. He worked and played on the circuit and then eventually, when I was visiting my mum, her boss had hired a new polo groom!

The rest as they say is history and I hope we have many more fabulous years ahead of us.

This sketch shows how as a child, Juan's father produced polo ponies to be shipped to the UK. He was a well-known polo pony producer and Juan was a good enough rider at a young age to help start any of the 'broncos'.

This isn't the place for a discussion on techniques or how horses are 'broken' worldwide. Yes, the Argentines can be firm and no, I don't condone it, but I understand why it happens.

The horses here are wild. I say here because as I type this, I am here in Argentina. They live naturally in herds, on the pampas, quite often never seeing many humans until they are three years old. Of course, by this age, they are bigger and stronger than when we would usually be starting to handle them and so the methods the humans use also become stronger.

Because Juan's family lived in a vast rural area, Juan rode his trusted pony the 5km to school and back, tying her to a tree so the neighbour would look after her until it was time to return home. Juan often jokes that was why he didn't have many girlfriends at primary school, because he always smelt of horses!

After polo, Juan started working in the racing industry and later working as an assistant stud groom, which meant he has delivered hundreds of foals over the last few years, putting his veterinary training to good use. I have also seen many of these births and I never get bored of seeing a new life come safely into the world. I just have reservations about what happens next.

I make no secret of my dislike of the racing industry. I cannot in any way, however you try to spin it, see that its ok to ride two-year-olds. Like many things though, disliking and complaining about something alone, doesn't bring about change.

To SEE a change, we must BE the change.

Some may say its hypocritical for my husband to work in an industry I clearly dislike. I don't see it like that. I know that the horses that Juan handles are treated fairly and kindly, and firmly where necessary but not cruelly, like so many of the horror stories I've heard.

He is leading by example, to be the change the horses in the racing industry so desperately need.

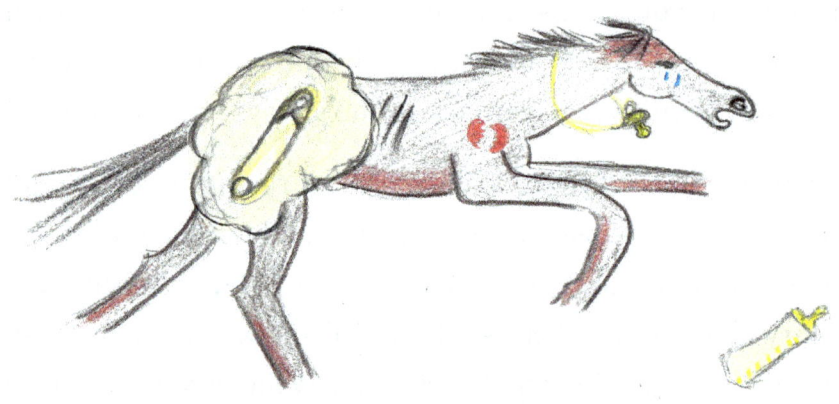

I really struggle with horses in industry because they are
mostly just babies. They are often rushed, have gaps in their
education, and are discarded when they are no longer
financially viable or profitable.

'You can never trust a horse who has been educated by fear. There will always be something they fear more than you. But, when he does trust you, he will ask you what to do when he is afraid.'

Unknown

To be welcomed into a South American family is a wonderful thing. The Latinos are as loving as they are funny, passionate as they are loyal, and as culturally different to us Brits as you can get.

There is no standing on ceremony; my house is your house, maté or a big bottle of beer are happily shared within the group as a sign of trust and friendship, everyone is welcome, and no one goes hungry!

Imagine the Italians, the French, the Germans, and the Spanish all mixed into one along with the indigenous Natives, and you have an incredible mix of history and cultures that our somewhat stuffy British culture might struggle to comprehend.

My family is a beautiful mix of skin tones and heritage, European and Indigenous bloodlines, that also includes an incredibly Spiritual side.

There are still many traditional healers practising old medicine, and although it's a culture cloaked in Catholicism, the old ways are still respected. Like many cultures there are folklores and stories going back decades.

Juan's paternal grandmother, Abuela Maria, was a Curar El Empacho, an old folklore practitioner who helped with indigestion or gastroenteritis healing, learnt from her mother-in-law, who was Indigenous to South America.

She used a head scarf twirled up into a ribbon, measured from her stomach to the person's using her elbow to her thumb along the line of the scarf, locating the problem area, and putting her fingertips in their diaphragm. Some skin pinching from the back and herbal tea also helped relive the symptoms!

When I was on Richard's Advanced Communication course, after one of the meditation exercises, I was told that there was an older native lady standing behind me, dancing with happiness that not only had I joined the family, but that I was pursuing my Spirituality and going into Animal Communication.

As I write this part of the book, I am enjoying some much-needed family time in Argentina; the parallel between my basic ability to communicate in a foreign language within my own species and how animals struggle to communicate with us, is not lost on me.

I'm lucky that all my family and the people I meet are patient and kind with me as I try my best to express my mood, my feelings, my thoughts ... and my behaviour is a result of that.

An incorrect tense, the wrong verb, it doesn't matter, as eager teachers await an eager student.

My basic tries are rewarded with kind corrections, an understanding and sympathetic ear, and my pleas to please slow down are immediately answered.

This all results in my confidence growing, my motivation to

learn increasing, and a desire to practice even harder.

If my attempts had been met with unkindness, disdain, impatience, and rudeness, there is a chance I would well and truly get the hump or just not bother anymore.

Both resulting in a decrease of confidence, motivation, and the downward spiral of isolation. Not hard to see the parallels with horses, is it?

Anyone who knows me, knows how difficult I find it to leave my animals. Holidays have always been stressful as I get anxious trusting other people with their care, and I experience separation anxiety when I'm away from them. I've since realised it was a control issue. The feeling of helplessness if they needed something and I wasn't there, what if something happened and I wasn't there? What if they think I've abandoned them? No-one else can look after them as well as I can! Through trust in Spirit and confidence in myself, those anxieties are virtually gone. Of course, I still feel a pang when I leave Orion these days, but now I can explain to him before I go, reassure him I will return, and check in with him while I'm away — and if he needs something, I can ask his Aunties to oblige! My recent trip to Argentina was a massive test of this and he was fine both physically and emotionally. All that stress just gone because of communication.

Translating Mental Imagery Communication Messages

Animals will use whatever means of communication they can to help us understand what they are wanting to say.

For me personally, these can come as mental images, physical feelings and emotions in my own body, tastes, sounds, and I can feel and hear words or phrases. This is also in conjunction with paying attention to their physical presence, their movements, their emotions, their tension, releases, and body placement.

Whilst my physical body mirrors theirs as much as possible with my arms becoming their fore legs, the images can take a little more interpretation and translation.

It's much easier now, as I've built up an incredible mental dictionary and the more animals I communicate with, the more fluid my translation has become. Like with any language, the more you practice, the easier it gets. It's all about thinking outside the box!

Each beautiful soul has so much to offer, and the amazing thing is that each one adds to my dictionary, ready to help the next animal, and so the cycle repeats as time goes on.

Of course, each animal is unique, but unfortunately a lot of the scenarios and circumstances they experience are not, and this is how common themes help us understand more quickly what they are wanting to convey.

The images I receive aren't always literal. They are symbolic and representational, and I often have to piece the jigsaw together with the help of their human.

I love doing this as it becomes a light bulb moment for many people, and they are often flabbergasted at how accurate the imagery is in relation to their circumstances. The human also feels very included in the process, which I have found is vital to help them make any changes necessary that may have been requested, and strengthens their bond together.

The one I use to explain this best is the badger.

When I have seen a badger in my mind's eye on horse visits, it hasn't always meant the animal. In many of the cases when I have seen one, it has referred to another horse called Badger.

When you think about it, if the horse showed me a bay, a chestnut, or a grey horse, I am still not much closer to identifying them, but a Badger takes me straight to the exact horse!

Of course, this may not always be the case, and it is not this image alone that is being used. It is in association with all the other feelings and sensations I will experience at the time.

This is why we must be aware of everything we are feeling and experiencing, and not just focus on one part, as we may miss the true intention behind the information.

An example of this is the mental image of a beautiful swan gliding majestically over a lake. On the face of it, this is an incredibly graceful picture, one of calm and serenity but, seeing this along with having a feeling of anxiety takes me to an outwardly calm exterior but a frantic scramble under the water, suggesting that they feel no-one has noticed their unseen turmoil.

Animal Communication changes your whole outlook on

everything, and once the subtleties become obvious, we soon learn to trust our instincts and not just our ears.

Animal Communication comes from our instincts, our heart, our mind, and our souls. When we learn to trust them, the floodgates start to open.

I only need to listen to how a human talks about their animal to know what the problem is. Unfortunately, our species has a lot to learn from so-called 'dumb animals'.

As animals ourselves, we really need to get a better handle on behaviour evolution, and leaving some traditions in the past.

Our arrogance as a species only highlights our ignorance as a species.

How we chase and worship little green pieces of paper and superiority doesn't contribute to the continuance of our species; quite the contrary, it does nothing but destroy it.

How humans measure success has a massive impact emotionally and physically on our horses.

Equestrian or Horseman. You decide.

Being a silent bystander and staying in our own lane is harder than it looks when there's so much evidence of how unhappy so many horses are, around us all. But we just hold steadfast and lead quietly by example and hope that by BEING the change, we will SEE a change.

Laura and Theo

Without Laura in my life, it would not be possible to do as much of this work, as she is Orion's babysitter!

It is a big responsibility looking after anyone's animals, but with a communicator's animals, they will tell tales if asked!

No such worries for me, as Orion is very happy with the excellent care she takes of him.

Darling Theo is so attached to Laura that he finds it difficult when she goes out without him sometimes. We had a long chat about it all and he understands now that she's always planning to come back.

A real mummy's boy but getting braver, little by little.

My dream is for science and Spirit to work together.

When Richard did Orion's communication, there were some physical niggles that Orion wanted addressed.

When the vet came, I said I didn't think he was 'quite right' in those areas and had them investigated. I wasn't surprised when science confirmed what Spirit had given, but I had been reluctant to say why I'd thought he wasn't quite right for fear of ridicule. This also happened with a physio I called in for Orion. She found all the same 'problem areas' that he had told Richard about.

Once I had done my training it was easy for me to be confident with Orion's information to scientifically qualified professionals, but I remembered how hesitant I'd felt before in confiding exactly why I thought there may be issues in specific areas.

As I started going out on visits, this same hesitation arose many times with my human clients. They were committed to the communication but were worried they wouldn't be taken seriously by the Science-based professionals.

I wanted to and needed to bridge the gap. I'm very careful about which professionals and services I recommend, I will only recommend those who I have used or would use myself.

I've always been incredibly fussy about who I will allow to have any contact with my horses, even more so now as I know how heavily trust features within those relationships.

I have seen some awful treatment of horses over the years from so-called professionals, and often wonder how on earth some humans allow their precious animals to be treated in such a way.

I remember when Grace was being treated at Newmarket and a very well-known and respected farrier was putting remedial shoes on her. Grace was still heavily sedated and wasn't very balanced. The farrier was rough with her and kept moaning that he wanted to go home and watch the cricket.

Seeing red, I marched into the vet's office, ready to explode. There was no point addressing him personally as I could see how little respect he had for Grace. I was furious and told the vet in no uncertain terms, that just because my Grace wasn't a million-pound racehorse, it didn't mean she shouldn't also receive the first-class treatment and respect I expected from them.

The stunned vet came back outside with me and oversaw the rest of the shoeing, much to the farrier's disgust!

Because I am so cautious about Orion's care, the people I surround myself with on his behalf are all those I would recommend anyway. They have all relished in my stories of communication experiences and I have since been able to recommend them to my own clients, as they have recommended me when they feel I may be able to help a particular partnership they have encountered.

And so, my list of 'Communication Considerate' practitioners was born.

A list of qualified science people who believe, or people who are open to my work, who are happy to take referrals when something has been highlighted during communication.

I have farriers, an osteopath, a saddle fitter, an equine iridologist, a sheath cleaner, a McTimoney animal chiropractor, an Intelligent Horsemanship professional, an essential oils practitioner, and a human well-being trainer and consultant.

Word also got back to me from a new Equine Veterinary Practice that my name kept cropping up amongst their clients. They had noticed an increase in their clients telling them I had seen their horses; the vet had followed up what had been highlighted during their communications and the issues were found to be accurate.

I have registered Orion at this practice, and it is such a joy to be able to confidently share his information without being ridiculed.

It also means that any common themes that arise such as horses, starting in their teens, commenting on the deterioration of their night vision, can go on to be investigated by specialised vets.

As I write this, my third year in business as an Animal Communicator is well underway, and I have so many people and animals to thank for so many wonderful experiences and for so much love shared.

To have a local vet taking clients seriously, without making them feel any kind of way for working with me, has been life-changing for so many horses and their humans.

To be able to approach a vet without fear of ridicule has encouraged so many more to come forward, resulting in horses getting treated earlier, quicker, and with more targeted healing.

We all want the same thing at the end of the day; to help the horse. And this is such a massive thing for horse welfare in our area.

Every single one is appreciated so much. We will make change in horse welfare if we just lead by example. I gifted this sketch to the vets' practice as a thank-you.

Here we see the dark clouds on the left and the empty sack of Ego and Baggage, leaving them in the past.

The clouds get brighter and are filled with love as they come closer to Spirit.

The vet has Spirit close by as the red areas of pain are identified by us both, working together,

I'm at the horse's head as we both now help the horse, together, hand in metaphorical hand.

Everything about Animal Communication comes from unconditional love and Spirit.

We must be prepared to work on ourselves, our own evolution and growth, in order to help others.

I have found my calling in this life and it's an absolute privilege to be a voice for all the animals.

I hope these sketches have given you a little more of an insight into the emotional and symbolic nature of Animal Communication and has given you cause to question why something is happening, rather than just dealing with the consequences.

Love, light, and Blessings to all of Mother Nature's creations from myself, Spirit, and all the animals.

Namaste